ALLELOPATHY
IN THE LIFE OF PLANTS AND THEIR COMMUNITIES

Andrei M. Grodzinskii

АКАДЕМИЯ НАУК УКРАИНСКОЙ ССР
ИНСТИТУТ БОТАНИКИ

А. М. ГРОДЗИНСКИЙ

АЛЛЕЛОПАТИЯ В ЖИЗНИ РАСТЕНИЙ И ИХ СООБЩЕСТВ

ОСНОВЫ ХИМИЧЕСКОГО ВЗАИМОДЕЙСТВИЯ РАСТЕНИЙ

КИЕВ НАУКОВА ДУМКА 1965

ALLELOPATHY IN THE LIFE OF PLANTS AND THEIR COMMUNITIES

Andrei M. Grodzinkii

Translation and Notes by

R.J. Willis
and
G.J. Rosinska

SCIENTIFIC PUBLISHERS

Published by:

SCIENTIFIC PUBLISHERS (INDIA)

Jodhpur	Delhi	Germany
–	–	–
5 A, New Pali Road *P.O. Box 91* *Jodhpur 342 001* *INDIA*	*4806/24, Ansari Road* *Daryaganj* *New Delhi - 110 002* *INDIA*	*Landstr, 240* *Frankfurt 60388 -* *GERMANY*

E-mail: info@scientificpub.com
Website: www.scientificpub.com

Print : 2021

ISBN: 978-81-7233-947-0
eISBN: 978-93-86102-04-1

Printed in India

АЛЛЕЛОПАТИЯ В ЖИЗНИ РАСТЕНИЙ И ИХ СООБЩЕСТВ

Allelopathy in the Life of Plants and Their Communities

CONTENTS

ABOUT THE TRANSLATORS

Dr. Rick Willis is a former Senior Lecturer in Botany, at the University of Melbourne, and is now retired. He has published widely on the topic of allelopathy, and is noted for his research on events and issues that have shaped the development of the science of allelopathy. His books include *The History of Allelopathy* (2005) and an English translation of Gerhard Grümmer's *Die gegenseitige Beeinflussung höherer Pflanzen - Allelopathie* published by Scientific Publishers in 2008. More recently his interests have extended into archaeology.

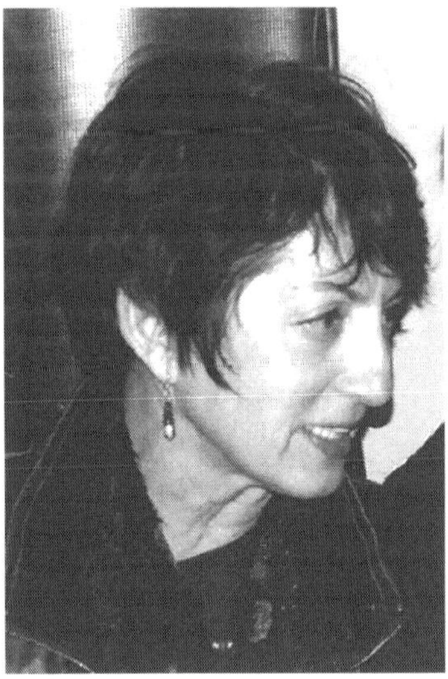

Grazyna Rosinska was born in Poland and studied Russian during her school and university years. In 1982 she migrated to Australia, and now works as the Arts Liaison Librarian at RMIT University in Melbourne, Victoria, Australia, where her interests include documenting imagery of contemporary art in South-east Asia

ACKNOWLEDGEMENTS

We are grateful for the assistance and cooperation of Dr. Anna Grodzinskaya and Tatiana Bugaenko, the daughters of Andrei M. Grodzinskii, and for supplying important resources and photographs. Also we thank Prof. Dmitrii M. Grodzinskii, the brother of Andrei Grodzinskii, for reading a draft of the biography. In addition, one of us (R.J.W.) visited the M.M. Grishko Central Botanical Garden of the National Ukrainian Academy of Sciences, Kiev in 1993, and was treated with the warmest hospitality at the Garden, at difficult economic times for the nascent Ukraine nation, by the Director, Prof. T.M. Cherevchenko, Dr. V. Grakhov, Dr. V. Kvasha, and in particular, the late Dr. Erast Golovko, who was then in charge of the Department of Allelopathy.

THE WORK OF ANDREI M. GRODZINSKII

R.J. Willis

Family History

The Grodzinskii family can trace its roots back at least to the 16th century, where one ancestor was a contemporary of Nicholas Copernicus at the University of Krakow in Poland. A continuing characteristic of the Grodzinskii lineage is education, integrity and independent thought. Apparently three Grodzinskii brothers fought in the Polish Uprising of 1830, and the one surviving brother was Andrei Grodzinskii's great-great-grandfather. Over the past two centuries, the family dispersed to Germany, Lithuania, Ukraine and Russia. Andrei Grodzinskii's ties to biology may be traced back to his grandfather, a noble who owned an estate near Lutsk, a city in northwestern present-day Ukraine, and who studied agriculture. Andrei's father was Michael Karpovich Grodzinskii (1890-1956), and his mother was Vera Filip Grodzinskaya (née Klimenko) (1907-1983), both of whom were involved with the Faculty of Agriculture at Bilotserkvivskyi Agricultural Institute, in Bila Tserkva, a city about 80 km south of Kiev. His father was one of the founders of the institute in 1920, and served as Dean for many years, and his mother became an assistant professor in Botany.

Childhood

Andrei Mikhailovich Grodzinskii was born on 3 December 1926, and his younger brother Dmitrii Mikhailovich Grodzinskii was born in 1929. The boys were exposed to natural history lore from an early age, which obviously served to direct their future careers, although it is regarded that Andrei wanted firstly to be a geologist, and had considerable interest in chemistry. Nonetheless, Andrei's father, following the revolutionary years was placed in charge of the large arboretum "Alexandria" in Bila Tserkva, a wonderful park and woodland, where Andrei, as a young boy, learned the formal names of a great many plants, and undoubtedly gained an irreplaceable sense of the purpose of natural botanical surrounds. Andrei's early childhood was full of pleasant memories, but the Grodzinskii family then had to survive years of misery, as in the period 1932-1933 the Soviet Union under Stalin attempted to subjugate the Ukraine through a program of planned mass starvation known as the "Holodomor", during which millions of rural Ukrainians perished through both starvation and disease. The Grodzinskii family, being linked to an urban agricultural institute, had access to basic food, although rationed.

Figure 1. Andrei as a young boy, holding in his hands a Russian "primer". (Courtesy of the family archives of Tatiana Bugaenko and Anna Grodzinskaya)

In 1941, during World War II, the Ukraine was invaded by Germany, which left Ukrainians faced with two evils – Russia under Stalin to the east and Germany under Hitler to the west. Ukraine suffered some of the worst Nazi atrocities seen during the war, including almost total annihilation of the Jewish population, and again seizure of food stocks, this time by the Germans. War events led to closure of the Bila Tserkva Agricultural Institute, and its evacuation of staff by the Soviets to the eastern, safer parts of Russia. The Grodzinskii family was evacuated eastward, firstly to Krasnodar in western Russia, where Andrei was able to continue school, then Krasnoarmeyesk near Moscow, and eventually to Alma-Ata (now Almaty) in Kazakhstan, about 3500 km east of Kiev. During the first months of leaving Bila Tserkva, the family, with Andrei's father seriously ill, was forced to find refuge in a rowboat on the Dnieper River near Cherkassy, as the area was being bombed. In Krasnodar, the family lived for many weeks in the open, and used their botanical knowledge to feed on wild berries, mushrooms, and abandoned crops such as potatoes and beets. The family eventually arrived in Alma-Ata, and Andrei became the provider for the family, and worked as a boiler attendant and technician at the Kazakh Agricultural Institute of Agriculture. However, food was in very limited supply and the family existed on a diet mainly of beetroot in various guises.

Education

In May 1943 the family was relocated once again, for the parents to resume teaching in Voronezh, which lies about 400 km south of Moscow, and which had been liberated from the Germans in January 1941. By late 1943, Soviet forces had reclaimed most of the Ukraine, and Andrei was able to enrol at Kon' Kolodyazkoho Agricultural College in the Lypensk district east of Kiev, and also obtain employment at the college. He graduated in 1944, but in March 1945 he was conscripted into the army, where he served with distinction for five years, and gained valuable practical skills in construction, especially during a term in Sebastopol, which were to serve well later in the development of the Central Botanical Garden complex in Kiev. During his army years, Andrei continued his studies via correspondence with the K.A.Timiryazyeva Agricultural Institute.

Figure 2. A young Andrei Grodzinskii in military uniform, 1948 (Courtesy of the family archives of Tatiana Bugaenko and Anna Grodzinskaya)

In 1950 he returned as a student to his home city of Bila Tserkva, where he studied at the Agricultural Institute until graduation in 1954. It is interesting that instead of sitting state exams, he chose the option of completing his studies by submitting a thesis entitled "Geomorphology, hydrology, soils and vegetation of the Bilotserkivskyi River floodplain area". He worked for a year as an agronomist at the Kozeletsk Meadow Research Station, and in 1955 he began his Ph.D. studies under the direction of Academician P.A. Vlasyuk at the Institute of Plant Physiology and Agricultural Research, Akademiya Nauk USSR, in Kiev. The topic of his research was the physiology of plants of the steppe, in particular in regard to photosynthesis and mineral nutrition. In 1958 he ably presented a defense of his thesis with the subject "Uptake and movement of nutrients in the plant under the influence of the treatment of soil and fertilisation", following which he received his Ph.D.

Family Life

It was following demobilisation and during Andrei's years as a tertiary student at Bila Tserkva that he met his wife, Galina Semenivna (née Zabigailo; born 1927), although they had actually known each other in primary school. The handsome couple was married in 1952, and their first child, a daughter, Tatiana, was born in 1954. A second daughter, Anna, was born in 1957. During this time, Galina remarkably was also engaged in research, and was doing her own doctoral studies on growth and development in sugar beet. Andrei remained closely associated with his younger brother Dmitrii, who became a plant physiologist, and a frequent collaborator in Andrei's research.

Figure 3 Andrei and Galina in 1951 (Courtesy of the family archives of Tatiana Bugaenko and Anna Grodzinskaya)

Professional Life

Towards the conclusion of his postgraduate studies in 1957, Grodzinskii was appointed as a junior research scientist in the Department of Physiology, Institute of Botany, Kiev, with the task of investigating the interrelations of steppe plants. It was in 1958 that Grodzinskii first became impressed with the concept that plants could affect other plants by means of chemical substances. Grodzinskii recorded that a researcher in his laboratory G.O. Kuznetsova, in the course of seed germination trials, had accidentally knocked a Petri dish containing seeds of *Crambe tataria* into another dish, where it caused severe inhibition. Subsequent research in Grodzinskii's laboratory led to his first publication concerning allelopathy: "Inhibitors of germination from fruits of Tatarian bread *Crambe tataria* Sebeök" (Grodzinskii *et al.* 1960, published in the *Ukrainskii Botanicheskii Zhurnal*).

Figure 4. Andrei Grodzinskii as a postgraduate student, with Academician P.A. Vlasyuk, *c.* 1952. (Courtesy of the family archives of Tatiana Bugaenko and Anna Grodzinskaya)

Grodzinskii served as a scientist at the Institute from 1957 until 1965, when he was appointed as Deputy Director, at this institute. During his tenure at the Institute of Botany, he expanded his interests in the chemical substances that act in the environment, and furthermore worked with brother Dmitrii in consolidating an understanding in plant physiology, which culminated with publication of "A Short Handbook of Plant Physiology" (A. Grodzinskii and D. Grodzinskii 1964), which became acknowledged as a standard work in its field.

Figure 5. Grodzinskii with German allelopathy pioneer, Dr. Gerhard Grümmer, Greifswald, DDR in 1963.
(Courtesy of the family archives of Tatiana Bugaenko and Anna Grodzinskaya)

Later in 1965, Grodzinskii was appointed as Director of the Central Republic Botanical Garden, Kiev[1]. His new position allowed him to establish a Department of Allelopathy, and he attracted scientists from different disciplines with interests in allelopathy: P.A. Moroz (horticulture), N.I. Prutenskaya (floriculture), N.N. Dzyubenko (agriculture), and G.G. Baranetskii (forestry). It was also in 1965 that Grodzinskii published his first comprehensive work on allelopathy, "Allelopathy in the Life of Plants and their Communities", a provocative work which was the accumulation of his research and thoughts on the matter over the previous ten years. Grodzinskii was awarded a D.Sc. for his efforts, and in 1967 was given the title of Professor by the Highest Certifying Commission of the USSR. It was also in 1967 that the Central Botanical Garden was officially recognised as a scientific research institute.

Figure 6. Grodzinskii at his desk, Central Botanical Garden, 1966.
(Courtesy of the family archives of Tatiana Bugaenko and Anna Grodzinskaya)

As the new director of the Central Botanical Garden, Grodzinskii had many responsibilities, including rebuilding much of the infrastructure, which had suffered decades of neglect, and his years of military service at Sebastapol gave him the confidence to take a leading role in design and construction of new glasshouses and facilities. Grodzinskii had a "hands-on" approach to his job, which perhaps gave him relief from the tedium of administration, and many of his staff and colleagues fondly remember him labouring outside from time to time, or recollect spotting a recently used scythe in his office. In 1967, he was appointed as Chairman of the Society of Botanical Gardens of Ukraine and Moldova.

Grodzinskii is chiefly celebrated for his pioneering work in allelopathy, although he fully acknowledged the contributions of other Soviet workers such as Sergei I. Chernobrivenko and N. G. Kholodnyi (both Ukrainian). During his prolific

---

[1] Today, this is known as the M.M. Grishko National Botanical Garden of the Ukrainian Academy of Sciences.

career, Grodzinskii authored or edited over 500 various botanical articles and works, of which the majority relate to allelopathy.

In 1967, Grodzinskii organised in Kiev the first All-union symposium on allelopathy, under the title of "The Physiological-Biochemical Bases of the Interactions of Plants", and the proceedings of this were published in 1970. This marked the "golden age" of an era concerned with the study of allelopathy, and in total from 1970-1975, six volumes of symposia proceedings were published under the title "Fiziologo-biokhimicheskie Osnovy Vzaimodeistviya Rastenii v Fittsenozakh" (=Physiological Biochemical Bases of the Interactions of Plants in Phytocenoses). Also during this period, Grodzinskii became increasingly interested in the topic of "phytoncides" or volatile emissions of plants that could be useful in therapy, antibiosis, ecology, etc., in conjunction with Boris P. Tokin (1900-1984), a Russian plant biologist with a long-standing interest in the topic.

Figure 7. Grodzinskii in a new conservatory, Central Botanical Garden, *c.* 1968. (Courtesy of the family archives of Tatiana Bugaenko and Anna Grodzinskaya)

In 1979 Grodzinskii established a department of Medical Botany at the Central Botanical Garden, and encouraged research on the psychological and health benefits of plants and secondary metabolites. He initiated a variety of programs, including research on the efficacy and bases of folk medicines, and innovations such as "phytoergonomics", that is incorporating living plants into built environments, where the visual, textural, physiological and aromatic (phytoncidal) properties of plants, were assessed in relation to human activity and well-being. Grodzinskii, besides promoting the therapeutic aspects of plants, envisaged plants as atmospheric filters that could recondition the air in a variety of situations ranging from crowded workspaces to factories to homes. For similar reasons, Grodzinskii was also involved in the early stages of designing plant experiments useful in the Soviet space program, and the Central Botanical Gardens provided mature epiphytic orchids and mycorrhizal orchid seedlings, which were sent into space with Soviet cosmonauts Popov and Ryumin on Soyuz 35 in 1980, to be grown on the space station Salyut 6.

Generally, the mature orchids did not fare well, and it is of interest that in the context of "excretion" in plants, it was concluded that plants suffered self-poisoning in the face of an apparent lack of gravity[2]. Later, Grodzinskii became interested in the idea of breeding plants with natural herbicidal potential, and in the concept of plant signaling and communication, discussed in an avant-garde paper, published in "Chemical Interactions in Plants" (1981), in which he presaged the idea of "phytosemeiotics".

Figure 8. Grodzinskii in the laboratory, with researchers, *c.* 1968.
(Courtesy of the family archives of Tatiana Bugaenko and Anna Grodzinskaya)

It must be remembered that during the early years of Grodinskii's directorship of the Central Botanical Garden in Kiev, relationships between the Soviet Union and the West were quite poor, during the so-called Cold War. However, despite these circumstances, Grodzinskii was able to travel for the first time to the United States to attend an international conference in March 1968 at Santa Barbara, California. This was fortuitous, as in August 1968, Soviet forces invaded Czechoslovakia, which likely would have been precluded Grodzinskii being allowed entry to the United States. The conference was part of the burgeoning American contribution to the International Biological Program (IBP), which had been in operation in Europe since 1964. The proceedings of the Santa Barbara conference were published in 1971 as "Biochemical Interactions Among Plants" and "Insect-Plant Interactions". As such, Grodzinskii was able to meet his American counterparts, Cornelius H. Muller and Elroy Rice. Grodzinskii attended conferences elsewhere, especially in the Soviet Union, East Germany, West Germany, Poland and Czechoslovakia. In 1985, under somewhat more favourable political conditions, Grodzinskii was able to return to the United States, where he attended the American Chemical Society symposium in Chicago on Allelochemicals.

---

[2] See p. 342 of: Harvey, B. and Zakutnyaya, O. 2011. *Russian Space Probes: Scientific Discoveries and Future Missions*. Springer, New York.

Figure 9. A.M. Grodzinskii in 1985, atop the Sears Tower in Chicago, where he was attending the American Chemical Society symposium on "Allelochemicals: Role in Agriculture and Forestry". The proceedings editor, George R. Waller, is behind Grodzinskii. (Courtesy of the family archives of Tatiana Bugaenko and Anna Grodzinskaya)

In 1973, Grodzinskii published his second monograph on allelopathy, and this, significantly, was written in Ukrainian, with the title "Principles of the Chemical Interactions of Plants". In this volume, he abandoned some of his earlier, tentative ideas, and presented an updated review of the literature and his own research.

Figure 10. Andrei with brother Dmitrii, 1983. (Courtesy of the family archives of Tatiana Bugaenko and Anna Grodzinskaya)

Grodzinskii, continued to conduct symposia on a regular basis, and monographs on allelopathy were published annually under his editorship. These include: "Problems of Allelopathy" (1976); "Allelopathy in Soil Sickness" (1978); "Interactions of Plants and Microorganisms in Phytocenoses" (1977); "Chemical Interactions in Plants" (1981); "Allelopathy in Natural and Artificial Phytocenoses" (1982) "The Role of Allelopathy in Aquatic Plants (1982); "The Role of Plant and Microbial Toxins in Allelopathy"; and "Experimental Allelopathy" (1987). Also, three volumes bearing his name were published posthumously: "Methodological Problems in Allelopathy" (1989), "Allelopathy and the Productivity of Plants" (1990), and "Allelopathy of Plants and Soil Exhaustion" (1991).

Figure 11. Andrei Grodzinskii, *c.* 1980. (Courtesy of the family archives of Tatiana Bugaenko and Anna Grodzinskaya)

During his years at the Central Botanical Garden in Kiev, Grodzinskii supervised some 30 Masters' students and 5 doctoral students, mostly devoted to studies relating to allelopathy. Grodzinskii was inordinately fond of talking to people about plants, ideas and possibilities.

Despite the initial good resources for the study of allelopathy in the Soviet Union, with Kiev as its focal point, there began to emerge significant opposition from more conservative elements, and research funds became increasingly difficult to obtain. It is interesting that a similar story unfolded in the United States and elsewhere, where from about 1964-1975, allelopathy was viewed as a new frontier; however, critics began to become dubious of the reality of allelopathy operating in nature, and established ecologists such as J.L. Harper seriously challenged the validity of allelopathic research. Grodzinskii remained as an optimistic driving force in allelopathic studies until his untimely death in 1988.

As mentioned earlier, despite Grodzinskii's preeminence in the field of allelopathy, his other contributions were also distinguished. He was keenly aware of the educational role of botanical gardens, and viewed such as essential to any city claiming to be cultured. His foundational work was vindicated in 1972 when the Central Botanical Garden was declared a site of national importance. Clearly, a major part of his brief as Director of the Central Botanical Garden was documentation of the flora of the Ukraine, and aiding conservation of rare species. Grodzinskii was also involved in a series of symposia and proceedings on "The Introduction and Acclimatisation of Plants in the Ukraine" from 1965 to 1983. Grodzinskii had been working for some time in organising material for a monograph on medicinal plants, and in 1992 the book "Medicinal Plants: an Encyclopaedic Handbook" was published, with Grodzinskii recognised posthumously as the editor. Grodzinskii also, in keeping with his role as Director of the Central Botanical Garden, became a great populariser of botany and science. He wrote a charming and award-winning book entitled "In Nature and in the Laboratory", edited a couple of science magazines, and hosted a popular television series about plants.

Figure 12. Grodzinskii amid *Crambe*, *c*. 1985.
(Courtesy of the family archives of Tatiana Bugaenko and Anna Grodzinskaya)

In 1967, Grodzinskii had been appointed as Deputy Secretary (Biology) of the Ukrainian Academy of Sciences, and in 1979 he was appointed as a full member (or Academician), and served as Academician Secretary of the Department for General Biology. From 1977 to 1984 he served as a member of the Council of Biological Sciences of the USSR. During his lifetime, he received many awards, including the Order of the Red Flag, the Badge of Honour, the Kholodnyi Prize, a silver medal from the Purkinje University in Slovakia, and the State Prize of Ukraine in Science and Technology (posthumous). A quaint honour was bestowed in 2009, when a street in Bila Tserkva was named in honour of Andrei Grodzinskii and his brother Dmitrii Grodzinskii. The International Allelopathy Society (IAS) has named its ongoing award for the best single publication on allelopathy after A.M. Grodzinskii.

Figure 13. Grodzinskii at his rotating desk in his office, hosting allelopathy scientists from India (from left, V. Rizvi, S.J.H. Rizvi and son J. Rizvi); V. Kvasha is on the right.

On 17 December 1988, Andrei Grodzinskii died aged only 62 years. He is survived today by his brother Dmitrii - a renowned plant physiologist and also Academician of the Ukrainian Academy of Sciences, his two daughters, Tatiana (Tanya) and Anna - a noted mycologist, and grandchildren who have continued the dynastic academic tradition of the family. His beloved wife Galina died in 1994.

Figure 14. A bronze memorial plaque honouring A.M. Grodzinskii was placed outside the Central Republic Botanical Gardens laboratory building, Kiev in October 1991. (Photograph by R.J. Willis, 1994)

Andrei Grodzinskii's life traversed the most difficult years of the 20th century, and in all of this, he remained foremost devoted to science and to service to his community. He lived during an era when technology was rapidly overtaking Nature, and both he and his brother were outspoken about the failings of government, in particular, disasters such as Chernobyl. It was perhaps the Chernobyl catastrophe that exacerbated the cancer that was to bring a premature end to his extraordinarily productive life. On 26 April 1986, one of the Chernobyl nuclear reactors exploded, spewing vast amounts of radioactive dust into the atmosphere, affecting hundreds of thousands of people in the Ukraine, Belarus and Russia. By early May, Kiev, which lay only about 100 km to the south, had become heavily contaminated, and Grodzinskii was involved in assessing the situation, and in dealing with contaminated plants at the Central Botanical Garden.

Andrei M. Grodzinskii is today acknowledged throughout the Ukraine and Russian speaking nations as their "Father of allelopathy". His office at the Botanical Garden remains untouched as a memorial, and the focal point is a very large round, custom-made table with a rotating top that helped him deal with his multitude of diverse tasks, Grodzinskii is fondly remembered, especially in his native Ukraine, where his integrity, initiative, erudition and scholarship are celebrated, and most recently a tribute volume was published by the Academy of Agricultural Sciences of the Ukraine in 2012, on the occasion of what would have been his 85th birthday.

Bibliography

Andriyevich, V. 2014. [The Grodzinskii dynasty of naturalists]. *Demokratichna Ukraina*, Tuesday 14 October 2014, pp. 4-5.

Davydenko, M.M. 2013. [Role of Academician A.M. Grodzinskii in the increase of level of scientific researches in the botanical gardens of Ukraine and Moldova]. *Visnyk Ahrarnoyi Istoriyi* 4-5: 269-272.

Davydenko, N., Bugaenko, T.A., Grodzinsky, G.A., and Grakhov, V.P. 2012. *Akademik AN URSR, Hrodzyns'kyy, Andriy Mykhaylovych (1926–1988) : Bibliohrafichyy Pokazhchyk*, do 85-richchya vid dnya narodzhennya (=Academician AN URSR: Grodzinskii, Andrei Mikhailovich (1926-1988) : bibliographic guide, on the 85 anniversary of his birth). *Vidomi vcheni-pryrodoznavtsi ta osvityany Ukrayiny* Kniga 5. Asotsiatsiya Bibliotek Ukrainy, Derzhavna Naukova Sil's'kohospodars'ka Biblioteka Natsional'na Akademiya Ahrarnykh Nauk Ukrainy, Kiev.

Golovko, E.A. and Afonina, E.A. 1986. *Andrei Mikhailovich Grodzinskii.* Naukova Dumka, Kiev.

Golovko, E.A., Grakhov, V.P., Boiko, E.N., and Bugaenko, T.A. 2006. *Andrei Mikhailovich Grodzinskii – Biobibliografiya.* Akademperiodika, Kiev.

Golovko, E.A. and Kvasha, V.V. 1991. Andrei Mikhailovich Grodzinskii (1926-1988). In: A.M. Grodzinskii: *Allelopatiya Rastenii i Pochvoutomlenie. Izbrannye Trudy.* Naukova Dumka, Kiev, pp. 5-20.

Golovko, E.A., Roshchina, V.V. and Narwal, S.S. 1995. Andrei M. Grodzinsky (1926-1988). *Allelopathy Journal* 2(1): iv, 1-7.

Grodzinskaya, A.A.[3] 2007. My father; glimpses of memory. In: *Botany and Mycology: Modern Horizons. To Memory of Academician A.M. Grodzinskii (1926-1988)* (ed. A.A. Sozinov). Akademperiodika, Kiev. Pp. 19-26.

Grodzinska, G. 2006. [A venturer in plant science: the 80th anniversary of the birth of Academician A.M. Grodzinsky (1926 - 1988)]. *Visnyk Natsional'noyi Akademiyi Nauk Ukrayiny* 2006(12): 54-62.

Kvasha, V. [c. 2013] *Rika Ditynstva i Yunosti - Diariush Maybutn'oho Akademika* [= Years of childhood and youth – diary of a future academician]. Privately published, Kiev.

Musatenko, L.I., Musiyenko, M.M. and Sytnik, K.M.. 2012 [Kiev Academician Andryi M. Grodzinsky – Truths of Allelopathy (devoted to 85th Anniversary of his birth)]. *Ukrainskii Botanicheskii Zhurnal* 69(1): 134-141.

[3] Note that Grodzinskii's second daughter Anna is known also as Ganna (or Hanna), and that the family name can be either Grodzinskaya or Grodzinska.

Translators' Note

The book *Аллелопатия в Жизни Растений и их Сообществ* [Allelopathy in the life of Plants and their Communities] was written by Andrei M. Grodzinskii and published in Russian by Naukova Dumka, Kiev, in 1965[1], and the full text is provided here in English, and indeed any foreign language, for the first time. Importantly, 2015 marks the 50th anniversary of publication of this seminal work.

In the modern era, there are only a few books devoted to allelopathy that precede the book by Grodzinskii. Molisch (1937) wrote a monograph that gave rise to the term "allelopathy"; however, this work mainly concerns the effects of the gas ethylene. In the period 1955-1957, three important books on allelopathy were published: Grümmer (1955), Chernobrivenko (1956), and a Canadian work in 1957 by Hubert Martin – of these only the work by Grümmer is reasonably well known, as it has been translated into Russian (Grümmer 1957) and English (Grümmer 2008). The pioneering work by Martin, "Chemical Aspects of Ecology in Relation to Agriculture" has been overlooked in most texts, including that by Grodzinskii, and even that by Elroy Rice, "Allelopathy", first published in 1974.

Grodzinskii's book is a foundational work that was in effect his doctoral thesis, and as such, it offers a remarkable, holistic overview of chemical plant ecology. It has long been neglected in the West for a variety of reasons. Firstly, it has been published only in Russian until now. Secondly, the book was published in a relatively small edition of 1700 copies in Kiev, in the former Soviet Union, and very few copies made it to the West; indeed the global catalogue WorldCat records only five copies: four in the USA, and one in Australia; although there are likely more copies in libraries of former Eastern bloc countries (e.g. Czech Republic, East Germany). It must be remembered that relations between the Soviet Union and western nations, especially the United States, were very poor at the time, as this period occurred during the height of the Cold War.

In translating this book, we have attempted to preserve as much as possible the style and format of the original book. However, there are always difficulties in translating a technical work from one language to another, especially where different scripts or alphabets are used. Where possible, the figures from the original Russian edition have been retained and modified with English text. As the book was written fifty years ago, there have been subsequent changes to things such as chemical terms,

---

[1] The book was republished in 1991 by Naukova Dumka, Kiev, as part of a commemorative volume of works by Grodzinskii entitled *Allelopatiya Rastenii i Pochvoutomlenie; Izbrannye Trudy* (=Allelopathy of plants and soil sickness; collected works).

plant names, plant family names, etc., and where it helps to clarify reading of the book, such changes have been made or noted. In a few cases, explanatory notes have been added to help amplify the text where the exact meaning may be unclear to the present-day reader. Explanatory footnotes supplied by the translators are those that are numbered; others with symbols are owing to Grodzinskii. Where possible, errors, ambiguities, and omissions have been amended.

There are a few idiosyncrasies in Grodzinskii's style of writing. Firstly, he sometimes writes in note-like form, and in such cases, we have used sentence structure. Also, as Grodzinskii is Ukrainian, he occasionally invokes Ukrainian words or terms. When citing works in the text, Grodzinskii commonly uses the author's first name or initial, in addition to the surname, and we decided to retain this style, albeit unorthodox today. In the original Russian edition, the references were divided into Cyrillic references, and references using western scripts. For the sake of convenience, the names of Russian and Ukrainian authors have been transliterated (using the "Modified Library of Congress System", as used by most American and British libraries), and merged with the other references in alphabetical order into a single reference list. The titles of Russian and Ukrainian works have been provided in translation, in order to provide more utility to the English reader. Also, we have decided to optimise the utility of the reference list by including missing details, most notably page numbers for journal articles, where possible.

Finally, it has to be said that both the translation and transliteration of Russian can cause some confusion. As indicated above, Grodzinskii was essentially a Ukrainian scientist, who at the time when the Ukraine was a republic within the Soviet Union, was compelled for political reasons to write primarily in Russian rather than Ukrainian. Even the author's name becomes greatly confusing, for it may appear in different westernised versions, depending on whether the source is Russian or Ukrainian, and on the system of transliteration. Currently, the preferred systems of transliteration do not use special characters. Nonetheless, Grodzinskii's name may be found variously in databases or bibliographies as: Grodzinsky, Grodzinskii, Grodsinskii, Grodzinskiy, Grodzinskiĭ, Hrodzinskii, Hrodzins'kyy, where notably the last two names are transliterations from Ukrainian, in which different conventions are used for certain shared Cyrllic characters.

ACADEMY OF SCIENCES UKRAINIAN SSR
INSTITUTE OF BOTANY

A. M. GRODZINSKII

ALLELOPATHY

IN THE LIFE OF PLANTS AND THEIR COMMUNITIES

THE BASES OF THE CHEMICAL INTERACTIONS OF PLANTS

KIEV 1965

The chemical interaction of plants by means of physiologically active excretions (allelopathy) is firstly discussed in this book as an independent biological problem. The role and place of allelopathy are shown in the other sciences, with its importance for plant physiology and biochemistry. geobotany, agriculture, forestry, and for the national economy.

Intended for scientific workers

Contributing Editor

Academician of the Academy of Sciences of the Ukrainian SSR, D, K Zerov

INTRODUCTION

The Program of the Communist Party for Soviet Biological Science established the target - "... the elucidation of the essence of the phenomena of life, unraveling the biological laws of the development of the organic world, the study of physics and the chemistry of life, the development of different ways of managing life processes ..."[1] and the need to address the intensification of chemicals used in agriculture, and it requires an in-depth development of the important fundamental problems of biology. One such problem is a longstanding, but little researched and controversial problem of the relationship between organisms. Examination of the issues of the mutual influences of plants, microorganisms and animals has great theoretical and practical significance. It is necessary to understand and control the process of evolution, to support a range of existing activities, and to develop new, agronomic, forestry and other activities; it is the basis of such scientific disciplines as geobotany, ecology, soil science, and others. It is due to the interaction of organisms that in nature, we know that there is a huge variety of life forms, from which breeders can draw material for their work. The interaction of organisms leads to strange evolutionary

---

[1] This edict originates from 1963 The CPSU Central Committee, USSR Council of Ministers in their Judgment of January 9, 1963 No. 63: "On measures for further development of biological science and strengthening its connection with the practice":

"In order to further successful development of biological science and strengthen its ties with the practice of the Central Committee of the CPSU and the USSR Council of Ministers decree:

1. Recognise the need to develop fully the whole complex of biological sciences. Considered the main tasks of these sciences are the elucidation of the essence of the phenomena of life, unraveling the biological laws of development of the organic world, the study of physics and the chemistry of life, the development of different ways of managing life processes, such as metabolism, heredity and direction of change in organisms. Need to develop broader and deeper Michurinian direction in biological science, which comes from the fact that living conditions are leading to the development of the organic world. Medical science should focus on the discovery of means to prevent and overcome diseases such as cancerous, viral, cardiovascular disease and other life-threatening diseases of humans. Theoretical foundations and practical measures to improve the longevity of people and their health should be developed. Equally important are the study and extensive use of microorganisms in the national economy and public health, including the development of food and feed, vitamins, antibiotics, enzymes, to find new methods of farming and animal science."

[Translators' note. Hereafter numbered notes are owing to the translators, and any other footnotes are in the original Russian text.]

events, but in most cases - appropriate harmony, suitable from the point of view of the progress of life on Earth, and from the point of view of maximising the use of living conditions.

Revelation of the lesser known aspects of the interaction of plants will be a new source for increasing the productivity of the cultivated and natural cenoses, creating sustainable and durable plants, and establishing measures to combat weeds and soil fatigue[2]. The problem of the interaction of plants has general biological significance, developed not only as part of a series of botanical disciplines, but it is closely intertwined with the problems of the relationship of the animals and microorganisms, and with soil science, geochemistry.

The study of the complex interrelations between plants growing together is attracting the increasing attention of researchers, which is the best sign of the scientific and practical importance of this problem. After many years of discussion regarding the problem of the formation of species, and after the publication of monographs by G. Grümmer (1955) and S.I. Chernobrivenko (1956), as well as numerous works by Soviet and foreign scientists regarding this problem, specific research areas have been outlined. It is becoming more and more clear that the problem of plant interrelations in its basis, deals with plant physiology, although this problem itself is generally biological and covers interests of general botany, geobotany, plant-growing, meadow cultivation, forestry, soil science, microbiology, etc. This is due to the fact that the diversity of relationships between plants growing nearby, is carried out mostly by physiological means, namely, by acting on the various life processes of plants, including their growth and heredity.

Nevertheless little attention has been paid to the problem of interaction in plant physiology, although among the pioneers of this problem are such famous physiologists as A. de Candolle (1832), H. Molisch (1937), P.S. Kostichev, N.G. Kholodnii (1957), J. Bonner (1951) and others.

Thanks to the work of N.A. Maksimov, A.L Kursanova and numerous Soviet plant physiologists who have reconstructed an overall picture of the physiology of the whole plant, there are coherent ideas about the relationship of physiological processes in the plant, based on the knowledge of the cycling of organic substances in the plant, the connection of all the parts and organs of plants, etc. And although many, many questions are still unclear and have only begun to be studied, the overall picture is already known and new results only complement or increasingly modify the picture, introducing new strokes or shades that do not significantly affect its overall shape.

Thoughts of physiologists are now directed to the interior of life – towards knowledge of intracellular life, towards studying the structure and functions of the smallest particles of living matter down to the molecular level. Research of these finest processes entails great difficulties and has become possible only because of modern technological and scientific achievements.

---

[2] Grodzinskii uses the Russian word *почвоутомление*, which translates as “soil fatigue”. This is a relatively non-specific term, known also as “soil sickness”, where there is a loss of production due to factors that are not easily alleviated by simple fertilisation.

At the same time, plant physiology should develop a broader scope of research. From the basis of the regularity of the physiology of the individual plant, we can move to studying the physiology of plant communities, physiology of plant interactions. In this direction, there is a lot of work, a lot of results obtained, and therefore, it has become necessary to generalise, and it is possible to draw a general picture that outlines the ways and methods of research.

The various manifestations of the interactions of higher plants in the community can be divided into two large groups. The first of them includes cases of cooperative use of the life factors by plants, in other words competition for nutrients, water, light, space, etc. One or a number of factors constantly appear to be in scarcity in any plant community, which leads to declining productivity of the whole community and changing of its species composition. Non-adapted plants are excluded from the community. It can be considered firmly established that plant interaction, on the basis of cooperative use of available resources, plays an exceptionally important role in the formation and development of natural phytocenoses[3], in the evolution of species and, in the development of highly stable, productive crops.

The cooperative use of any factor should not be considered as a disadvantage. In the case of surplus of a factor in the environment, some decrease of it because of its use by a neighbour plant has a positive effect.

This type of interaction is closely connected to the chemical interaction of plants or allelopathy. It is commonly known that plants secrete into the environment many organic physiologically active compounds. Dead plant mass, decomposed by heterotrophic organisms, is an inexhaustible source of other physiologically active compounds. This creates a definable biochemical environment around the plant, a distinctive protective sphere made of the active substances, that have a positive or negative influence on other organisms. In fact, some free organic substance in some form is present in any environment, where a plant grows. This substance is a cooperative product of vital functions of the higher and the lower autotrophic plants and different heterotrophic organisms; it continuously changes, and gets destroyed and supplemented by new secretions. Almost always this substance is not indifferent in biological relationships, influencing plants and microorganisms harmfully or positively. We do not know the chemical composition of very many substances, found in plant secretions, or in their transformations in the soil. At this stage, it is important for us to solve basic allelopathic problems.

The principles of allelopathic plant interaction, substances and environmental factors, taking part in this process are shown in Figure 1.

[3] East European plant ecologists have a lexikon that frequently differs from that in the West. A "community" is often referred to as a "cenosis"; a plant community as a "phytocensosis"; and a "biogeocenosis" is comparable to an ecosystem, etc.

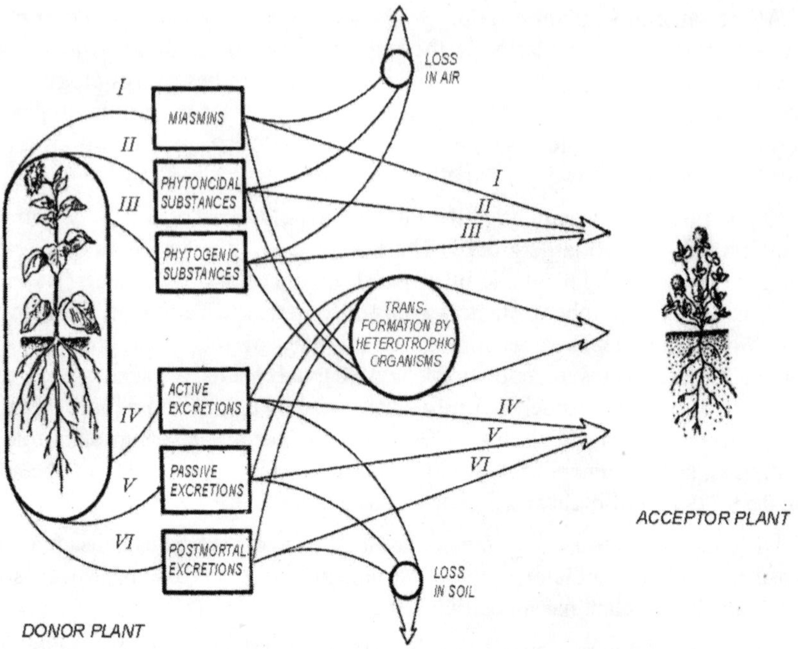

Figure 1. Scheme of allelopathy

A donor is a plant, of which secretions[4] we are discussing, it is analogous to the German term *Wirtpflanze*. An acceptor is a plant, that is found under the influence of the secretions being studied, it is analogous to the German *Gastpflanze*. Obviously, in a pair of plants each of them is simultaneously donor as well as acceptor. Evidently, the donor plant's secretions can reach an acceptor plant in gas form or in a water-soluble state. Hence two types of substances, water-soluble and volatile are important in allelopathy. Volatile substances are those that can exist as a gas, steam, aerosol (mist) or dust in the biological range of temperature..

We suggest to divide plant secretions into **intravital** and **posthumous**, that appear as a result of dying of the organism or its separate parts. Intravital secretions in their turn can be divided into **active**, resulting from normal plant vital functions, and **passive**, caused for example by rain washing out substances from leaves, etc. According to this, we divide volatile secretions into three types (Figure 1): **phytogenins** (Kholodnii, 1957; Sanadze, 1961), i.e. active, partly passive secretions from undamaged plant organs; **phytoncides** – secretions from the damaged tissues resulting from a disturbance of intact glandular substances, and autolysis (in a slightly more narrow sense than, according to B. P. Tokin, 1960); **miasmins** – secretions from dead, decomposing tissues (from the Greek miasmos – foulness).

---

[4] The term *выделение* is one of the most common words that appears in the Russian text of the book. In English, it can have many meanings, and this will depend on the context. Generally the word in botany refers to any sort of outflow of material from a plant, and thus the word can translate necessarily, as discharge, excretion, secretion, exudation, emission, etc.

Water-soluble secretions also can be divided into three types: active – **exudates**, passive – **diffusates**, and posthumous – **saprolins** (from the Greek sapros – rotten).*

On the way from the donor plant, secretions can be partly absorbed back by the plant that has produced them, partly undergo different transformations under the influence of abiotic factors (light, oxygen, etc.) or heterotrophic organisms. As a result of this, the secretions can be formed into new active products or fully mineralised, and finally a part of the organic secretions can reach the neighbouring plant.

So the task of allelopathy as a scientific problem, is to study the formation of physiologically active substances in plants, secretion by these plants into the environment, the transformation of secretions on their way from one partner to another, and, finally, the influence of active substances on the acceptor plant. Following G. Grümmer (1957), we call these substances cholines; sometimes "inhibitors" is a synonym for cholines, although they will only have an inhibiting, toxic influence in certain concentration, and if diluted, they normally stimulate the vital functions. This conditional terminology should be used until better terms are suggested.

In other words, allelopathy can be defined as the **cycling of physiologically active substances in the biogeocensis**. The beginning and the end of this cycling – formation of substances in plants, their absorption and influence on other plants – belongs to the realm of plant physiology and biochemistry; the middle belongs to soil biochemistry and soil microbiology.

As a whole, the problem of allelopathy is a part of the phytocenology and the study of evolutionary theory, and as will be shown later, can have diverse theoretical and practical usage in the different areas of science and production.

Chemical interaction is discussed in monographs by S.I. Chernobrivenko (1956) and G. Grümmer (1957). Since their publication, hundreds of works offering important new ideas and facts have appeared. These works have been mostly published abroad. There is some lack of generalisation and critical analysis of the released material.

This book is a result of many years of research by the author and his colleagues at the Institute of Botany, Ukrainian Academy of Sciences, and a generalisation of a significant amount of published works by Soviet and foreign scientists. The book does not claim the full elucidation of all chemical interaction problems, because many of them have only started to be explained. Wide research in this area, that has begun in the Soviet Union and abroad, promises us many unexpected discoveries, bright observations, valuable facts and conclusions for practical usage.

As at present, the ideas of allelopathy are not shared by all the scientists working in the areas of plant interaction; many biologists, specialists in agriculture and forestry are unfamiliar with this problem at all. Hence, very many facts remain unnoticed, many phenomena are interpreted incorrectly, etc. It is appropriate to quote

---

* Similarly the German is *Rottesubstanze* – decay substance

J. Liebig from his letter to F. Wohler[5]: "I am studying a new problem about the origin of ideas in natural science. I have found in the history of natural science as follows: in order to understand facts one should have in the head some ideas. Thousands of people see a phenomenon which owes its cause to a particular idea." (Letter of 29 June 1865).

---

[5] The passage originates from *Aus Justus Leibig's und Friedrich Wöhler's Briefwechsel in den Jahren 1829-1873*, published in 1888 (vol. 2, pp. 193-194), but is contracted and thus somwhat distorted. The letter to Wohler actually refers him to a lecture that Liebig was planning to give at the Royal Academy of Sciences in Munich in July 1866 in honour of the birthday of the Bavarian king, Ludwig II, and subsequently published as a pamphlet under the title "Die Enwicklung der Ursprung der Ideen der Naturwissenschaft" in 1866, more or less as a sequel to his "Induction und Deduction" of 1865. Liebig argues, as have many philosophers before him, that concepts and facts must precede principles and theories, and this differentiates inductive and deductive science. The quote in full is: "The philosophers assume that indeed facts generate ideas, and in the history of natural science, that in order to understand a fact one must have some ideas already in the head, and that the eyes see nothing, that has not been thought in the mind before. Thousands of people see a phenomenon occurring, without thereby any idea being aroused in them, and if the numbers so decide, it must be concluded that the facts do not generate ideas per se; a thought is but created in the mind , and it seems clear to me that this must be a conclusion. Whatever man does not think, neither acts according to his senses, and if we have a false theory in mind, we thus make the contradictory phenomenon of no impression, we record it or put it in the bed of Procrustes."

Chapter 1

HISTORY OF RESEARCH AND PATHS IN THE DEVELOPMENT OF THE PROBLEM OF CHEMICAL INTERACTION OF PLANTS

The interaction between plants has been observed by man since time immemorial. Even when trying to cultivate plants, the ancient farmer was faced with the impossibility of growing together many plants in the same area, with the failure of field cultivation of plants under the canopy of the forest, with the rapid loss of fertility with ongoing cultivation, etc. These circumstances were one of the sources of tillage systems. In order to overcome the resistance of wild or ruderal vegetation and allow the growing of useful plants, the soil must be treated, sometimes by spending a great deal of effort to eliminate the spread of weeds, overgrowth of forest, etc. At various times, depending on the economic, geographical and other conditions, this problem was solved in different ways. In the ancient times, slash and burn agriculture arose intuitively, as men in order to bring new land into cultivation, resorted to burning forest, shrubs or turf. According to A.V. Sovetov (1950), "burned forests was called **ляд**[1], brush – **сыросек**[2], and turf - **кубыш**[3]. The common name for such cultures is "**fireplace**" or **burns**. Burning dry grass of the last year is often used in steppe conditions. The core of the fire system was to destroy interfering plants - trees, shrubs, turf, root shoots, etc. and their seeds, - as well as to accelerate mineralisation of organic matter present in the cenosis in order to ensure a good growth of cultivated plants. With "burned forest" or "burns" a crop was harvested several times, after which it was abandoned.

In the south, in the steppe zone, there arose in a similar way a shifting or fallow farming system. In this case, the land received light surface treatment for several years and was seeded with crops as necessary, and after exhaustion ran to fallow (deposit), which was used as pasture or for hay-mowing. Then the land was used once again for ploughng. With this system, weed control was effected by natural steppe formation of the land and displacement of weeds, and mineralisation of organic matter was achieved by ploughing and harrowing the soil.

---

1 The Russian word *ляд* is likely a local term that refers to the hellish or unclean nature of such places.

2 *Сыросек* is an old word that refers to burned steppe areas.

3 Assumedly this is a local reference to a region in southern Russia characterised by soddy vegetation.

With increases in population density and land becoming private property, extensive opportunities for extensive land use decreased. The result was a three-field system of agriculture, which consisted in the alternation of fallow sites (tied) with winter and spring crops. This system arose spontaneously from the shifting system as a result of its reduction over time. More intensive use of the land in a three-field system required its fertilsation. Depending on the economic and geographic conditions and traditions, this problem was solved in different ways: applied manure, ash, or grass cultivation (known as Norfolk rotation). Neither the ancient nor the newer systems were able to provide the highest standards of farming.

The three-field system and remnants of the shifting system survived in Russia until the October Revolution. With the development of large-scale occupation of the land, tenure based on capitalist principles, there arose new, so-called crop rotation systems of cultivation. These pursued as their main purpose maximising the potential fertility of soil, rapid recovery of fertility through the use of various fertilisers and intensive farming techniques, diversification of cultivated plants by introducing a large number of industrial crops. Thus arose scientific agronomy, theories of crop rotation, and the theory of mineral nutrition, which provided the basis of these farming systems.

Without dwelling on the detailed characterisation of all these systems, we note only that the intensive exploitation of land has led many countries to enormous loss of productive areas due to water and wind erosion, destruction of the structure, "burnt out" humus, etc. Especially a lot of arable land was destroyed by rapacious farming in the United States of America.

Grassland agriculture, developed by V.R. Williams, was recommended for implementation in the Soviet Union in the years of collectivisation and after that. The essential feature of this system was a stake in the natural regeneration of soil fertility, where quite large areas were set aside for sowing of perennial grasses. The grassland system did not take into account the enormous possibilities of modern mechanisation and use of chemicals in agriculture, insofar as at the beginning of collectivisation, the country experienced a huge shortage of tractors, machinery, fertiliser, labor, etc. It may be to some extent justified then that with the development of powerful industrial base of the U.S.S.R, the agricultural grassland system began to hinder further growth of the productivity of the fields.

Therefore, quite timely and very relevant was the question of the development of new farming systems, based on a thorough understanding of the general laws of nature, and accounting for specific soil and climatic factors and economic conditions.

We have done this brief overview of agricultural systems in order to show the importance of the interaction of plants in agronomy. All pre-existing and existing systems of agriculture were intended to overcome soil sickness, overcoming the impossibility of continuous cultivation of one and the same crop on the same areas, overcoming weeds and other harmful vegetation, etc., with integrally producing maximal output at the lowest labour cost. In ancient times, the former requirement was accomplished by frequent changes of cultivated areas; now it is due to the intensification of cultivation of the soil. However, it is clear that soil sickness, comprises the total or partial loss of soil fertility. The harm of weeds, in terms of decreased yield and deterioration of its quality, is a direct or indirect manifestation of the interaction of plants. A farming system should provide accelerated circulation of substances in the cenosis: to get the maximum yield from cultivation, it should break down organic matter created by previous generations of plants, as carbon and mineral salts are bound in inaccessible form to this matter. It must be broken down, but not completely - as a certain amount of organic matter should remain in the soil.

Allelopathy, as the cycling of physiologically active substances in the cenosis, has direct relevance to farming systems. Namely, an excess of physiologically active organic substances in the environment of the cenosis is harmful for the growth of plants, as is also a shortage. In all farming systems, tillage of the soil regulates the level of cholines, while farmers did not even realise this.

From the above, it is clear that the study of the interaction of plants in space and time is crucial for the theoretical study and the development of farming systems. From this follows the first direction in the study of the interaction of plants, which arose a long time ago and which could be termed agronomic. There are the well-known views of Theophrastus, Columella, Pliny the Elder about the importance of the interaction of plants. The centuries-old practice of farming has gained a very large amount of factual material about the possibilities of co-cultivation of plants, density of sowing and planting, weed control, crop rotation, and many others.

This direction has been developed mostly empirically. Only in the last decade, have there been attempts experimentally to isolate the individual factors involved in the interaction of agricultural plants. Among these factors differentiated are competition for water, light, nutrients and specific chemical effects.

A very valuable summary of the earlier studies of the chemical interactions of plants and a large amount of original factual material are presented in the book by S.I. Chernobrivenko "The Biological Role of Plant Secretions and Interspecific Relations in Mixed Crops" (1956). The author distinguishes between mutual or unidirectional, favoring or inhibiting growth, and indifferent types. From the possible combinations of these influences, he developed the scheme, shown in Figure 2.

The big drawback to studies on the chemical interaction of crops is that the excretions of plants are seldom ascertained. Their activity has been confirmed by various indirect indicators; however, it has not been convincing enough to prove the existence of chemical interactions.

The significance of allelopathy for growing agricultural crops is also revealed in the works of A. G. Winter (1955, 1960), G. Grümmer (1955, 1961 a, b), H. Börner (1956 a, b, 1958, a, b, c and 1960 b). B. Rademacher (1958), J. Bonner (1950), H. Cézard (1960), Z. Laštuvka (1955, a, b, c, 1960), A., M. Groszinskii *et al.* (1961), K. G. Lavrova (1961) and others. Closely related to the study of chemical interaction in crops are numerous works devoted to the soil humic substances, which are recently summarised in monographs by Kononova (1963) and A. Kloke (1963), as well as research on the biochemistry of agricultural soils (Peive, 1961) and the relationship of higher plants and microorganisms (Krasil'nikov, 1958).

Despite a number of positive indications relating to allelopathic effects of certain crops (rye, flax, chickpea, buckwheat, yellow lupine, guayule) and weeds (wormwood, couch grass, false flax), the chemical interaction of plants is totally inadequately taken into account in agricultural science and practice. Growers and agronomists explain the interaction of plants in fields only from the standpoint of competition for light, water, nutrients, the generation of specific microflora, the accumulation of pathogens and pests, the deterioration of soil physical properties, etc., almost without paying any attention to the accumulation of physiologically active substances in the environment.

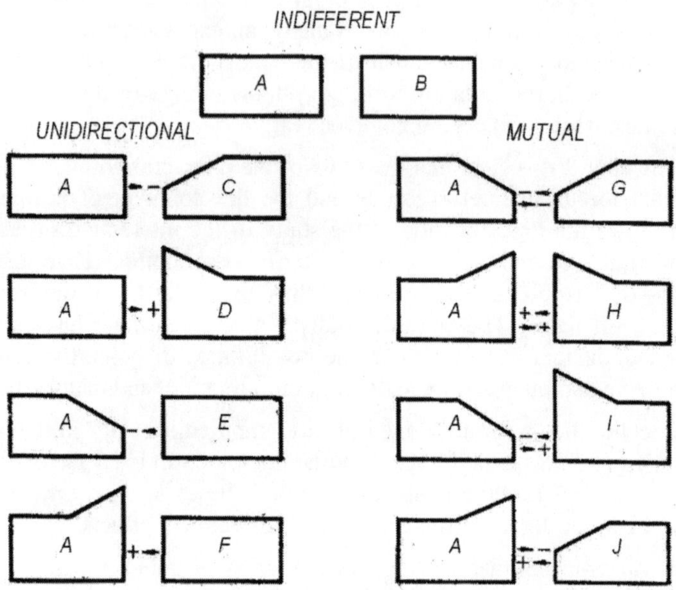

Figure 2. Scheme of biological interactions among plant species (from S.I. Chernobrivenko, 1956)

Meanwhile, in many long-standing practices, the methods are based on the unconscious use of some aspect of allelopathy. Beneficial effects of various kinds of organic fertilisers - manure, humus, green manure, compost, faeces, and others - depend not only on that they contain mineral nutrients and organic products that support the livelihoods of the microflora, but also on the physiologically active compounds that have a direct impact on the growth of higher plants. It is never possible to repeat the effect of an organic fertiliser, using equivalent amounts of mineral nutrients. Adding food for microbes, such as carbohydrates (straw) leads to the binding of biological nitrogen and even phosphorus, that impairs plant nutrition. Hence a significant part of the effect of organic fertilisers must be attributed to the physiologically active substances - cholines.

The classical successful combination of oats with vetch is based not only on the improvement of the nitrogen nutrition of the former, but in similar mixture with peas, the oats is inhibited. In numerous experiments in recent years, combinations of corn with different leguminous plants have given very contradictory results, although in all cases the corn grew above the legumes and was not shaded by them, as it is better able to tolerate the lack of moisture, etc., so that competition is not the case, where the legume retards the growth of corn.

The harm caused by weeds cannot always be clearly explained as due to their competition, although of course, it plays a very important role. Completely inexplicable from the standpoint of competition, is the increased growth of crops in the presence of a small admixture of weeds, which has been described and noted by many researchers and practitioners.

Insufficient consideration of the particular features of the chemical interaction of plants can result in some damage in field cropping, vegetable and greenhouse beds, hydroponics, horticulture, afforestation, grass growth, ornamental floriculture, etc.

The second important trend in the study of the interaction of plants growing together originated and developed later. It refers to the characteristics of the interaction of plants in natural phytocenoses - forest, meadow, steppe, etc. and originates with the beginnings of the scientific concepts of vegetation. This trend can be called botanical. In its development of great importance were the works of prominent botanist-naturalists: A. Humboldt, A. de Candolle, and at a later time – the geobotanists V.N. Sukachev, A.Ya. Gordyagina, A.P. Shennikov, M.V. Markov , E.M. Lavrenko, T.N. Rabotnov, B.A. Bykov, Clements, J. Braun-Blanquet (1961), R. Knapp (1954, 1960a), A. Scamoni (1955), L. Guyot (1951, 1959) and many others.

Botanical understanding of the interaction of plants formed the basis of phytocenology - the teachings of the laws of vegetational cover on Earth and its individual components - phytocenoses.

The most progressive, in our opinion, are the views advanced by V.N. Sukachev (1942, 1945, 1957) and developed in the works of Soviet geobotanists (Lavrenko, 1959, Markov, 1958, 1962, 1964). At the heart of the concepts they put forward are ideas about the unity of the organism and the environment, the relationships of flora and fauna, including the world of microorganisms, and the role of plants in shaping the environment. Based on the provisions put forward by V.I. Bernadsky, Sukachev developed the concept of the biogeocoenosis as an interconnected and interdependent set of all living beings inhabiting a certain area of terrain and abiogenic factors operating within the area.

Understanding the complex nature of biogeocenoses emphasises the necessity for their multifaceted study, "The main objective of these composite studies is to be able to get closer to a clear understanding of the process of circulation and transformation of matter and energy among the components of the biogeocoenosis, and between them and other biogeocenoses, and in general with other phenomena of nature." (Sukachev, 1957).

Lavrenko (1959), in setting out ways of plants interacting in ecosystems, divides them into direct (immediate) and indirect relationships. The first group includes parasitic, symbiotic, physiological (fusion of roots), biochemical (allelopathy), mechanical (strangling lianas "girdling" branches) and epiphytic relationships. Among the direct relationships of most phytocenotic importance are biochemical, symbiotic, and in part, parasitic relationships. Indirect relationships are universal; they are divided into habitat-forming and competitive.

Lavrenko acknowledges that biochemical effects play an important role in the life of plant communities, because they affect the relationships between plant species, and in some cases, completely eliminate the possibility of the development of certain types of plant communities in having a definite composition. It is very likely that in some cases, single-species or few species communities are united not so much by the competitive power of the complete dominance of a species, i.e., its ability to actively use the environmental conditions, but by its excretions, that have a negative effect on many other kinds of plants. "Unfortunately, - the author notes, - with all these biochemical effects, especially in particular plant communities, study is still woefully inadequate."

Foreign phytosociologists partly accept the classification of the relationship between plants, built on the principles put forward by V.N. Sukachev (Scamoni, 1955); many also adhere to the views of the Swiss school of geobotany formulated in the famous textbook by Braun-Blanquet (1951). Accordingly these relationships are divided into those of connection by close dependence (e.g. symbiosis, parasitism), and commensalism (from the French. commensal), i.e., connection by plants sharing common sources of life factors. Chemical interaction is neglected quite frequently in the work of this school.

More attention is given to the chemical interaction of plants in the book by R. Knapp, "Sociology of Higher Plants" (Knapp, 1954). Based on data in the literature and his own experiments on seed germination of some plants in the presence of other types of seeds or extracts from different parts of plants, the author writes quite confidently about the part of allelopathy in interactions in communities. However, the conclusions of this book are not always convincing, since one cannot transfer the results of laboratory experiments to nature. R. Knapp does not provide evidence to support the presence of chemical interactions in specific phytocenoses.

In geobotany long dominated by descriptive research methods, the introduction of experimental methods encounters great difficulties. In only superficially observing and recording certain facts of good or bad growth, the retreat or advance of certain groups, etc., it is difficult to judge the true causes of the manifestation of these facts. Therefore, the easiest explanation invokes conventional, proven, familiar, well-known reasons, which may in fact be wrong. A lack of proper representations hinders the development of the science.

Let us briefly examine the history of the emergence and development of the field of study of allelopathy.

The existence of chemical interaction, as previously noted, is even mentioned by botanists of antiquity (Cato *et al.*, 1937).

The idea of chemical interaction was especially popular at the beginning of the last century, during the reign of the humus theory of plant nutrition as advanced by A. Thaer and K. Sprengel. Photosynthesis at the time had already been discovered, but not until the works of de Saussure and Boussingault was any attention paid to it, and it was believed that plants take their organic matter from the soil.

Therefore, the task of the landholder was the full enrichment of soil humus. On this basis, no-one doubted the possibility of mutual influence of the plant through secretions.

For the first time a clear concept of the mutual influence of plants was formulated by A. de Candolle (1832), who believed that the substances excreted by plants can cause significant harm to the same species and to other species of plants. On the basis of these views, A. de Candolle developed the theory of crop rotation, aimed at overcoming soil sickness, and a theory of plant association. Both of these theories have played a very significant role in the development of scientific agronomy and in the formation of geobotany as a science.

Later, the humus theory was abandoned with the advent of the brilliant works of Liebig and his numerous disciples and followers, who created the theory of mineral nutrition of plants and showed the principle of the possibility of growing plants in a fully mineral medium. However, J. von Liebig was not as critically opposed to humus theory, as is often attributed to him. He believed that in this theory, there is a valuable kernel, and in his later years he welcomed the attempt of Grandeau[4] to combine the two theories, which, however, was unsuccessful.

The views of A. de Candolle have long remained unnoticed because of the dominance of the theory of Liebig, that all the facts of the interaction of plants can be explained only from the standpoint of competition, the unilateral removal of substances, etc.

Only at the end of last century and the early years of this century did research resume on the mutual influence of plants through their excretions. Reasons for this were the discovery of substances with a high physiological activity - enzymes, vitamins, and the need for practical resolution of the issue of soil sickness. From this period are the works of S.U. Pickering, Schreiner *et al.*, which are discussed in detail by S.I. Chernobrivenko (1956) and G. Grümmer (1957a).

Of great value for the development of the problem of the chemical interaction of plants was the famous physiologist H. Molisch (1937), who coined the term allelopathy (from the Greek. αλλήλωυ - mutual and πάυή - influence).

The work of academician N.G. Kholodnyi (1957) on the volatile secretions of plants and soil microorganisms, conducted by him during the period from 1938 to 1952, opened a new dimension of the problem and outlined interesting ways to resolve it. He also suggested a number of original research methods to study volatile phytogenic substances from the atmosphere. His ideas were echoed in the studies G.A. Sanadze (1959, 1961), D.J. Persidsky and Wilde (1954), G.V. Porutskii *et al.* (1962, 1963) and in the work of the Allelopathy Laboratory of the Institute of Botany of the Ukrainian SSR.

In 1950 there was a review article by J. Bonner, in which he summarised the matters conducted in association with Went, Galston and Gray in studies of the root exudates of guayule (*Parthenium argentatum* A. Gray) and secretions from the leaves of the desert shrub encelia (*Encelia farinosa*).

The literature on the chemical interaction of plants up to 1953, mainly foreign, was summarised by Grümmer (1955, 1957a), who proposed the following scheme of the mutual influences of plants:

---

[4] Louis Grandeau (1834-1911) was a French chemist and agronomist, who published several works on agricultural chemistry.

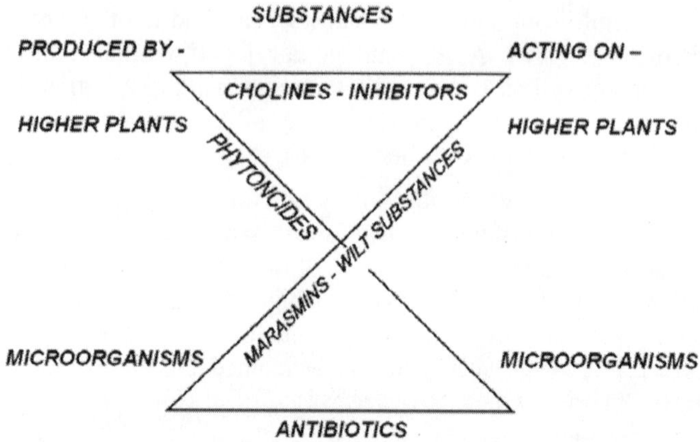

Antibiotics are meant as substances that are secreted by microorganisms and act on other microorganisms. According to the proposal of Gaumann and Jaag (1946) marasmins (from the Greek, marasmos - weakening) are products of metabolism of parasitic microorganisms that cause pathological effects on higher plants (in particular wilting, necrosis). Phytoncides, here, are substances that act on microorganisms. For substances produced in higher plants and that have a toxic effect on other plants, G. Grümmer coined the term "cholines".

In this scheme, the author makes an attempt to divide the substance according to their origin and mode of action. Actually, allelopathy refers only to the top line - cholines - showing the direct effect of active secretions among the plants. Other possible ways of chemical interference, mediated through the flora and fauna, are not considered in the scheme. The disadvantage of this scheme is that it only shows the negative mutual influence of organisms.

M. Kh. Chailakhyan (1962) extended this scheme by introducing into it the notion of not only inhibitory, suppressive substances, but also that of metabolic products of higher plants and microorganisms stimulatory to growth, as chemical interaction involves not only depression, but also the stimulation of growth.

The proposed scheme under consideration separates physiologically active substances that are not always strictly distinguishable by the function of the individual substances. Thus, the same substance may be a choline and a volatile product; depending on the concentration, it can have also positive and negative effects, etc. In his scheme M. Kh. Chailakhyan used simultaneously biological and chemical nomenclature of compounds.

B. Rademacher (1959) instead of "cholines" offers the neutral expression "allelopathica", i.e. substances which, depending on the concentration, have a negative or positive allelopathic effects.

G. Grümmer (1957b, 1961b) pays great attention to chemical interaction of field crops and weeds (flax and false flax, couch grass).

The German school of allelopathy researchers has produced dozens of excellent works. Among the first of these studies should be mentioned that of H. R. Bode (1940, 1958) on the influence of volatile secretions of wormwood on fennel

(*Foeniculum vulgare* Mill.) and the allelopathic properties of walnut and black walnut (*Juglans regia* L., *J. nigra* L.). The current study of the allelopathic and biochemical properties of substances of *Artemisia* continues with H. Schwär (1962 a, b, 1963).

Many studies have been made by A. G. Winter, his students and colleagues, who studied the presence of antiphytotic and antibacterial substances in various types of organic residues in the soil. They expressed and substantiated the hypothesis that allelopathic action concerns substances of phenolic nature. Recently, this group has paid great attention to the nutrition of plants via organic compounds.

B. Rademacher (1959) published a review article on the mutual influence of the higher plants, and as did H. Börner (1960) on the role of toxic excretions in soil sickness.

Some interest in elucidating the role of chemical interaction in natural phytocenoses, relating to the excretion by roots and leaves of sugars and amino acids, the effect of various secretions and extracts on the growth of plants, etc. appears in the works of G. F. Linskens and R. Knapp,

The works of V. Flaig, E. Zalbach, F. Scheffer and others are very important for the development of allelopathy, understood as the circulation of physiologically active substances in the biogeocenosis, for they focus on the decomposition of soil organic matter and the formation of physiologically active humus-like substances. An interesting review of the literature on this issue and much original experimental data are presented by A. Kloke, (1963).

In France, the issues of chemical interaction of plants involved the geobotanists L. Guyot, G. Becker, M. Massenot, J. Montague, G. Deleuil, H. Cézard and others. Their works are of value in the sense that they are carried out in conditions of natural cenoses and were aimed at solving specific theoretical or practical issues. Good reviews concerning the physiologically active excretion of plants were presented by P. Duquenois (1955, 1958) and R. Cézard (1961).

In-depth studies of the chemical influence of couch grass on wheat and other plants, and the interaction of wheat and rye in mixed sowings were performed at Brno by Z. Laštuvka and his staff. By providing an integrated, comprehensive approach, they were able to establish a number of new biochemical, anatomical, physiological patterns; indirect allelopathy was shown: the concentrations of excretions that have no apparent effect on the growth of wheat do reduce its resistance against fungal diseases.

A few works appear occasionally among the publications of American scientists (Bonner, 1950, Holm, 1964), Swedish (Osvald, 1949, Aamisepp, Osvald, 1962), Australian (Cannon *et al.*, 1962), Japanese (Takijima, 1963), Indian (Sirtsar, 1961 a, b). Not all of these are devoted specifically to the mutual chemical influence of organisms, but they contain interesting facts and guidelines for allelopathy.

In the Soviet Union, in addition to the above-mentioned works by Kholodnyi, much research has been conducted by S.I. Chernobrivenko (1956), which deals with the interaction of plants for over 20 years. Studies by S.I. Chernobrivenko and his students (Gulyaev, 1963) are concerned mainly with the interaction of plants in mixtures, especially the interaction of cereals and legumes. These studies, apart from their theoretical interest, are of great practical importance, as mixed cultures in production do not always succeed, but the reasons for this are still unknown.

Great interest from the standpoint of chemical interaction studies is the production of phytoncides. On the issue of phytoncides, which was formulated by B.P. Tokin in the early 1930's, there have been hundreds of papers published, All-Union Conferences held, and the accumulation of a wealth of factual material. Phytoncides are increasingly used in medicine, veterinary practice, and plant growing. In considering phytoncides as a device for chemical protection of plants from unwanted living neighbors - microorganisms, protozoa, insects, we at the same time recognise that they are cholines and take part in the chemical interaction of plants.

In particular, much of our understanding about the role of phytoncides in the chemical interaction of plants has been contributed by the Leningrad botanist A.A. Chasovennaya (1954-1961). The physiology of the formation and release of volatile phytoncides and their biochemical properties have been studied by S. Skvortsov (1960, 1964).

The problem of phytoncide production is closely related to the study of antimicrobial substances in the plant world (Drobotko *et al.*, 1959). The presence of antimicrobial substances in plant excretions and plant debris is of utmost phytocenotic importance.

The accumulation and transformation of physiologically active substances in soils is the subject of the book by Ya.V. Peive, "Biochemistry of Soil" (1961). N.A. Krasil'nikov (1958), and N.E. Gvamichava (1962, 1963) studied the accumulation of vitamins in soils, A.V. Manorik and V.F. Vasilchenko (1964) studied the accumulation of vitamins and other physiologically active substances in manure and compost.

In his works, K.E. Ovcharov (1961, 1964) discusses the role of vitamins in allelopathy, particularly in the relationship between higher plants and micro-organisms, and between microorganisms.

The study, over a lifetime, of the volatile secretions of plants has been carried out in Tbilisi by G. A. Sanadze, V.V. Chubinidze V. *et al.* With very delicate and sensitive equipment, they have secured many entirely new facts that are important for understanding, not only the secretory function of plants, but also photosynthesis.

Original work on the exchange between plants, of labeled nutrients, has been conducted by S. S. Shain *et al.* (1957, 1959, 1962), G.V. Nikolaev (1963), N.V. Timofeev-Resovskii (1961), and L.D. Rachkov (1963). Of particular interest for the chemical interaction of plants is research on the exchange between plants of the labeled products of photosynthesis, which has been started by V. P. Ivanov.

The overall interaction of forest species has been studied by I.N. Rakhteenko (1963), but in terms of allelopathy, by M. Kolesnichenko (1957-1963), T.I. Golomedova (1952, 1954), and others. This research is increasing every year; it is closely linked with that of other sciences and is getting ever closer to suit immediate practical use, which is the goal of every scientific field. Soviet science in the field of allelopathy occupies a leading position and has a number of indisputable achievements.

Existing work on allelopathy suggests that the phenomenon of chemical interaction between plants can be considered completely and reliably proven. Not only *in vitro*, but also in the nature, plants are able to influence each other chemically. However, in the works of particular authors, there is a large disparity in the definition and scope of the problem of allelopathy. Very often, there is no clarity in the approach to the study of chemical interference, and almost no standard methodology. There are especially large disagreements about the evaluation of the biological role of allelopathy. While some of the researchers maintain that there are but few cases of allelopathy, that in particular, qualify as examples, others reduce all the interactions between plants to allelopathy. However, science has already accumulated enough facts, more or less, to resolve satisfactorily all of the existing doubts.

The subject of allelopathy is the study of the formation and release of active substances, accumulation and transformation of these substances in the environment, and their actions on other plants in cenoses in relation to environmental, agronomic, and other conditions.

To successfully resolve this problem, simultaneous research at different levels and at different stages is needed. For the convenience of study, the whole problem can be divided into four parts:

(1) accumulation and allocation of physiologically active substances of plants (sub-problems of donor plants);

(2) transformation and accumulation of active substances in the environment (environmental sub-problems);

(3) uptake and physiological effect of active substances on plants (sub-problems of acceptor plants);

(4) the overall effects of interactions (sub-problems of the practical application of allelopathy).

In conclusion to this chapter, it again should be stressed that the sources of the physiologically active substances in phytocenoses are as diverse and limitless, as the inexhaustible photosynthetic activity of green plants. Allelopathy is part of the eternal and continuous cycling of organic matter in the biosphere; at times there occurs excessive accumulation of cholines, saprolins, miasmins – at other times, almost their complete disappearance. Fluctuations in the levels of active substances in different structural elements of phytocenoses - daily, seasonal, and throughout their development cycles - are completely unknown to us. There is still much to be done jointly by scientists before this page of Nature will reveal its secrets.

Chapter 2

EXCRETORY FUNCTIONS OF PLANTS. THE PLANT AS A DONOR OF ACTIVE SUBSTANCES

The term "plant excretion" is used in the literature to refer to a discharge in the true sense of the word, for example, guttation, manna, nectar, root exudates (Eberhardt, 1954, 1955), the secretion of essential oils, etc., as well as various kinds of diffusates, solutions, and volatile vapors, etc., derived from the root system of dying, fallen leaves, stems, bark, flowers, seeds, etc. Excreted substances are those washed away by rain from living leaves, or substances that are released as a result of damage to the plants - phytoncides, honeydew, gums, resins, etc. (Kirillov, 1964).

All of the excretions of plants, that form the biochemical protective environment around the plants that produce them, are important in allelopathy.

In speaking of discharges or excretions, we assume that it should be the refuse, waste, by-products of the vital functions of plants. However, in many cases, this is apparently not so. Among the plant discharges identified are many that are useful and important for the plants themselves that produce them. The presence of vitamins and enzymes within the excreta, and the very common occurrence of absorption of recently excreted substances, further strengthens our opinion in this matter. Indeed, many of the excretory substances - oxalates, essential oils, resins, have been shown in anatomical studies, or by using the method of labeled atoms, as being involved in metabolism.

One might think, therefore, that the biological purpose of excretory functions in plants is different from that of animals, which, as they have great need of energy materials, pass huge amounts of organic matter through their bodies. In autotrophic plants, the energy needs are supplied in excess by photosynthesis. The development of large leaf and root surfaces in plants inevitably entails a loss in the space components of living protoplasm - solid, liquid and gaseous substances. An apt expression by Timiryazev is "necessary evil"[1], in reference to transpiration, which can be extended to other cases of the acquisition and expenditure of substances.

---

[1] Kliment A. Timiryazev (1843-1920) was an important and progressive Russian botanist and plant physiologist, who wrote widely, and of which his best known book is *Zhizn' Rastenij*, published in English in 1912 as "The Life of the Plant".

The special excretory activity of plants should not be underestimated. This was pointed out by P.S. Kostychev (1933), who called for the establishment of a new chapter on plant excretions in plant physiology.

Based on the classic definition of the essence of life given by F. Engels[2], as well as from the rich factual material, it can be argued that the plants always actively secrete into the environment considerable amounts of various waste products. Even during active growth and the accumulation of organic matter, the plant gives back to the environment a lot of different compounds. For example, G. Michael and H. Marschner (1958), using the tracer method, showed that the absorption and excretion of phosphorus occurs simultaneously in roots in aqueous culture, but with differing intensities. On average, in their experiments, the excretion of phosphorus is ⅓ of the amount of that absorbed in the same time. In experiments, D.V. Fedorovskii (1958) showed that corn, during 20-50 days in soil culture, excretes 4-25% of the phosphorus absorbed during the same time. Excretion is not dependent on the soil type, as it is osmotically weaker. Similar results were obtained by M. Kutaček *et al.* (1956) with wheat seedlings in water culture. The same is true of other cations and anions. Some time ago, K. Luttkus and R. Bötticher (1939) showed that potassium is absorbed in light (during daytime), and at night it is released back to the environment; the amount of loss can be 15-30% of the total amount of potassium in the plant. The same was shown under field conditions by J.F. Dubovik (1950) for calcium, chloride and sulphate ions.

As for the amount of active organic excretions as compared with the assimilation (photosynthesis) of organic substances, such data are essentially absent. It is possible to prove by indirect calculations that organic secretions make up a significant amount of the matter. Soil respiration, mainly due to the microflora, is on average 7.5 million dm^3 of CO_2 per hectare per year, which corresponds to the biological combustion of 10000 kg of sugar per hectare.

Since the higher plants are the only producers of organic matter (except for the small amount from soil algae, lichens or photosynthetic bacteria), and as the roots residues in the case of fairly high yield are equivalent to about 6000-7000 kg/ha of carbohydrates, the remaining 3000-4000 kg/ ha is the balance owing to active root exudates, decomposed by the microflora to carbon dioxide, and substances used in respiration by the roots themselves. According to data obtained by G. Lundegårdh (1957), the respiration of roots in the soil is not more than 1/3 of respiration of microorganisms, i.e. it may be the equivalent of about 1000 kg/ha of sugar. Thus active water-soluble organic excretion through the roots can constitute 2000-3000 kg/ha.

V.T. Tranquillini (1959) conducted experiments in a mountain forest in Austria, where he determined the annual balance of CO_2 assimilation and loss in young trees of stone pine (*Pinus cembra* L.)[3]. It was found that during the growing season, the

---

[2] Frederick Engels in 1894 gave a surprisingly succinct and mechanistic definition: "Life is the existence form of proteic structures, and this existence form consists essentially in the constant self-renewal of the chemical components of these structures".

[3] Grodzinskii refers to *Pinus cembra* as *кедр,* which generally translates as "cedar", but is also a Ukrainian common name for stone pine.

needles of trees assimilated 7830 mg CO_2 per 1 g of weight. During the year, of this amount, 2993 mg CO_2, i.e. 38% of the assimilation was lost as a result of respiration, and 4837 mg was useable for the synthesis of organic substances. This amount would be enough to build about 2.2 g of material per 1 g of needles. In fact, production of organic matter averaged only 0.65 g (ranged from 0.54 to 0.90 g per 1 g needles), i.e. three times less than potential. V. Tranquillini explains this deficit as due to possible error in the technique of measuring photosynthesis and in the calculations, and finally the loss of organic matter in the environment by secretions, which are used by mycorrhiza. The latter factor is that which the author considers of the greater importance. Thus, these experiments convincingly demonstrate that plant excretion makes up a very significant amount compared to the overall plant productivity.

Such discrepancies between the calculated value of net assimilation and the actual growth of the vegetative part of the plant are often found in other studies, but they have not been analysed, and they have been explained only in terms of the imperfections of the methods used in the study of photosynthesis.

There are also other methods of calculation, based on, for example, the weight of the soil microflora and renewal rate of microbial cells. According to N. A. Krasil'nikov, the arable layer contains 6-8 t/ha of bacteria, fungi, actinomycetes and algae, which during the growing season there are 10-20 successive generations, i.e. during this time they can synthesise 60000-160000 kg/ha of bacterial organic matter. Even if one considers the fact that there are long food chains, and microorganisms consume each other successively, so that the above figures are much exaggerated, the amount of organic matter, nonetheless, needed for the maintenance of vital processes of 6-8 t of microflora should be significant.

In general, it is not an exaggeration to say that the annual production of all plant secretions is under normal circumstances several tonnes per hectare. Thus, over the lifetime, the mass of excretions is of the same order as that of the passive release – from dying plant parts, fallen leaves, etc.

Since active organic secretions are not taken into account when assessing the photosynthetic productivity of plants, and they account for a substantial amount of our understanding of the assimilative capacity of the plant, then productivity is apparently reduced by 1.5-2 times. This should attract the attention of biologists concerned with photosynthesis.

Inhibitors[4] of germination of fruits and seeds and their role in the distribution of plants

The existence of substances in the fruits and seeds of plants that delay germination was realised long ago (Aksentzhev, 1927). Practitioners know that to enhance the

---

[4] The term *тормозители* is translated here mostly as "inhibitors", as commonly in English, one speaks very generally about substances that inhibit germination. The Russian word has a different nuance, as it refers to the process of retarding or braking. The inhibition of seed germination is a complex process that involves many different types of substances, and processes, such as dormancy, stratification, embryo growth, root and shoot growth and elongation, etc., and thus the single term "inhibition" is perhaps too simplistic.

germination of seeds in many plants, they needed to carefully free them from residual fruit or rinse them with water. For example, washing with water greatly accelerates germination and improves seed germination in oriental plane (*Platanus orientalis* L.). In an experiment without rinsing, seed germination was 7%, washed once in 24 hours - 23%, washed twice - 32%.

The seeds inside fleshy fruits (pumpkins, melons, cucumbers, tomatoes, lemon, etc.) germinate in the fruit in very rare cases, despite the fact that they have excellent germination once they are removed from the fruit, and the conditions of humidity and even aeration inside the fruit are usually satisfactory for germination.

The accumulation in the seeds and fruits of substances that can inhibit seed germination, and the growth of roots and stems is known for many species of plants. The data on this issue are contained in the works of M. Evenari (1949, E. Niemann (1952), Toole *et al.*, 1956).

Moreover, I.N. Isip (1938, 1939) suggested that germination inhibitors play a protective role in the relationship of plants with other organisms. Further observations have confirmed the validity of this assumption (Svinarev, Barannikova, 1956; Grodzinskii *et al.*, 1960; Lesik, Zviglyanich, 1961).

The release of substances occurs purely passively during swelling; nonviable seeds can release into the environment the same such substances, as do as viable seeds, and even in larger quantities. However, it does not preclude the possibility of a mechanism involved in regulating the release of substances by changing the permeability of living cells.

According to our observations, when seeds swell in water, they may lose up to 2.5% or more of their initial weight. Detected in extracts are sugars, many amino acids (Börner, 1955, 1956b), mineral compounds such as phosphates (Blaim, 1955), and other substances. Inhibitors of germination are found in extracts from the whole seeds of many plants.

The influence of external conditions on the release of substances by seeds has been little studied. Obviously, all of the factors that increase the hydration of proteins, the solubility of various substances, and the permeability of protoplasm contribute to the loss of substances from seeds.

The excretion of substances from seeds largely depends on their biological properties. As a rule, cold-resistant maize varieties release a smaller amount of substances during swelling than do non-cold hardy varieties (Segeta, 1963)[5], and the so-called "hard" seeds of legumes less than normal seeds (Primushko, 1963). Damage to the testa of the seed, or soaking it in warm water accelerates the loss of substances. The metabolites of microorganisms, as demonstrated by our definition, also accelerate the release of substances from seed (Figures 3 and 4).

---

[5] Grodzinskii attended the conference on the "Physiology, Ecology and Biochemistry of Germination", 8-14 September 1963, held in Greifswald, East Germany. Accordingly, he cites here, without publication details, the papers by Segeta, Mohill and Paleg, and Kutaček, although these were not published formally until 1967, well after publication of Grodinskii's book. While something of an anachronism, we have provided for these authors the references to their papers as published in 1967.

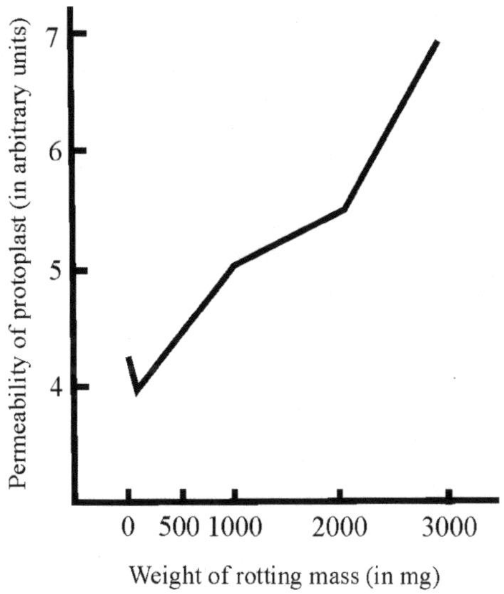

Figure 3. The effect of increasing concentrations of marasmins on the excretion of substances by swelling seeds of corn

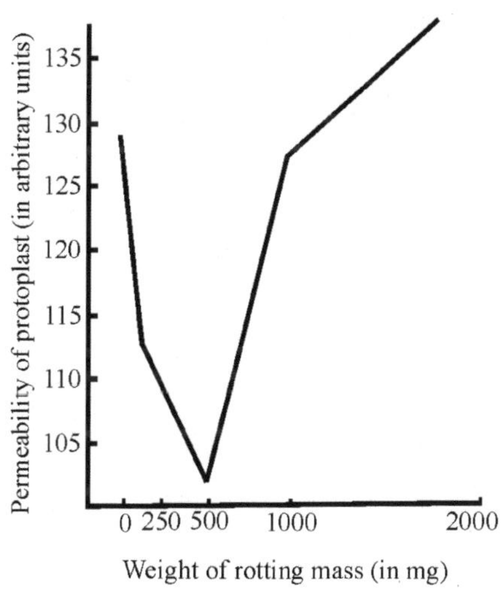

Figure 4. The effect of increasing concentrations of marasmins on the excretion of substances by swelling seeds of peas.

The method of study inhibitors of seeds and fruits usually involves the preparation of aqueous extracts and testing their effects on seed germination or growth of other plants in bioassays. The seeds of wheat, lettuce, radish, mustard, cress root growth or flax are those most often used. We have tested the inhibitory activity of extracts or volatile substances on the germination of radish (see Appendix).

Among the inhibitors identified in seeds are di- and tricarbonyl organic acids (Peers, Daniels, 1957), amino acids (Blagoveshchenskii, 1953, Kolobkova and Kudryashov, 1960), unsaturated aromatic acids - caffeic, cinnamic and ferulic acids

(Bonner, Galston, 1944), aldehydes - citral, cinnamic aldehyde, salicylaldehyde, benzaldehyde (Toole *et al.*, 1956), and unsaturated lactones - coumarin, scopoletin (Blaim 1955), protoanemonin, parasorbic acid, ammonia (Stout, Tolman, 1941) and flavonoid pigments (Novotelnov, Ezhov 1956), dehydroacetic acid, and desacetyl phthalide (Toole *et al.*, 1956), indoleacetic acid (Segeta, 1963).

Inhibitors in the fruit and seeds are often found in bound form, so light hydrolysis liberates significant amounts of active substances. As an example, in Table. 1 are the data from the extraction of fruit inhibitors from Tatarian bread[6] (*Crambe tataria* Sebeök.), which was carried out by infusion of samples of dry fruitlets in a solvent 10-fold by quantity, overnight at room temperature.

Acetic acid and ammonia are used because they are easy to remove by drying, and they do not interfere with the bioassay. Acetic acid not only extracted a lot of inhibitors, but it also increased the amount extracted a second time with water from the same seed. By contrast, ammonia yielded less leachable inhibitors than pure water, and also reduced the extraction when repeated with water.

Table 1. Extraction of inhibitors from fruits of Tatarian bread with acid and alkaline solvents (content of inhibitors as RCU).

Solution	First extraction	Repeat extraction (water)
Water	250	183
2% acetic acid	110	585
2% ammonia	202	78

Table 2. Release of inhibitors of germination by seeds (fruits) of tanning knotweed - *Polygonum coriarum* Grig. (content of inhibitors in aqueous extracts 1:10 as RCU)

Experimental variable	Origin of seeds	
	Tashkent 1959	Kiev 1959
Dried seeds	28.2	10.2
Stratified seeds	9.5	8.0
Germinated seeds after stratification	85.0	36.0

Note: Experiment was in 1960

[6] This is more or less a local name for *Crambe tataria*, which otherwise is known as Tatarian colewort.

During stratification, the amount of inhibitors released from seeds decreases, and at the time of germination it increases dramatically. This is because of the improved access of water due to the deterioration of the testa and regeneration of inhibitors due to increased hydrolytic processes (Table 2).

We studied the inhibitory soluble substances that are extracted from the fruits or seeds of plants when swollen in water. Seeds or fruits were not crushed, but were used in the form as they occur in nature. In total we analysed 2700 samples characterising the 827 plant species belonging to 62 families, common in Ukraine.

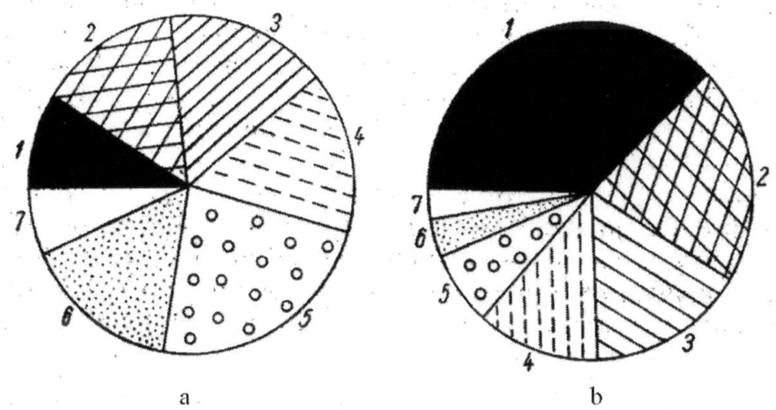

Figure 5. Occurrence of inhibitors in extracts from seeds (a), and from fruits and pericarps (b)
1 – toxicity; 2 – strong inhibition; 3 – average suppression; 4 – moderate suppression; 5 – weak suppression; 6 – no effect; 7 – stimulation in bioassays

The processed results of these tests are presented as pie charts in Figure 5.

The presence of inhibitors is typical for all the families that have been investigated. The highest average content of inhibitors are those characterised from some members of the buttercup family (*Adonis wolgensis* Stev., *A vernalis* L., *Paeonia tenuifolia* L.), Rosaceae (*Prunus mahaleb*, *Rosa canina* L.), many plants of the families Legiminosae[7], Malvaceae, Euphorbiaceae, Umbelliferae, Solanaceae, Scrophulariaceae, Cruciferae, Labiatae, Cucurbitaceae, Compositae, Chenopodiaceae, Salicaceae, and others; the least - pine seeds, flax seeds and fruits, primroses, birch, beech, Convolvulaceae, cereals, amaranth, lilies.

[7] Grodzinskii generally provides plant family names in Cyrillic, which means that they can be translated variously. Plant families are now generally standardised, with Lain names bearing the ending "*-aceae*", and being based on a type genus. However, there are eight plant families which have long held alternative irregular names, which by convention are still recognised: Leguminosae = Fabaceae, Umbelliferae = Apiaceae, Cruciferae = Brassicaceae, Labiatae = Lamiaceae, Compositae = Asteraceae, Gramineae = Poaceae; Guttiferae = Clusiaceae; and Palmae = Arecaceae. As Grodzinskii's book was written 50 years ago, it is likely that these older names would have been used.

A general relationship between the accumulation of inhibitors and systematic position of plants cannot be established.

Usually, the plants that are known to possess a high content of secondary materials (essential oils, tannins, alkaloids, pigments, etc.) in the fruit, or are toxic, or have medicinal properties, are also characterised by the presence of inhibitors of germination. However, these features are not always correlated.

We have confirmed the presence of inhibitors of germination in seeds and fruit, as shown previously in the literature, for example, in the seeds of the Siberian peashrub[8] (Kolobkova, Kudryashov, 1956), the balls of sugar beet (Maksimovich, 1957), parts of the fruit of pumpkin (Molotkovskii, 1949), etc. However, these inhibitors are not the strongest. There are found in many plants (*Crambe*, *Adonis*, *Cucumis*, etc.), fruits with exceptionally strong inhibitors of germination.

It would be difficult to expect that the presence of water-soluble inhibitors be closely linked to the systematic position of plants, since it is not an individual matter, but a very complex biochemical phenomenon. One might think that this feature is more of a physiological and ecological adaptation and it depends on the conditions in which the plant grows constantly, and the method of dispersal of the seeds.

The water-soluble inhibitors of seeds and fruits of various plants vary quite considerably. In the large majority of plants, content of inhibitors in the seeds themselves is significantly less than in the parts of the pericarp, sometimes by a factor of tens or hundreds. Conjecture about such patterns in the distribution of inhibitors has been expressed by E. Niemann (1952). However, families such as the Ranunculaceae, Onagraceae, Fagaceae, Lentibulariaceae, Compositae and Liliaceae (?) are exceptions in this respect. In our studies of the representatives of these families, seeds contain more or the same amount of inhibitory substances as is in the fruit pericarps.

The distribution of inhibitors in the parts of the fruit is usually such that the least amount of inhibitors is in the seeds, as we move away from them, the amount of inhibitory substances increases (Tables 3 and 4).

Table 3. Content of inhibitors (in RCU) in aqueous extracts at 1:10 from oats (*Avena sativa*) and wild oats (*Avena fatua*)

Sample	Oats	Wild oats
Cleaned grain	8.1	17.9
Spikelet glumes	34.0	44.0
Floral glumes	230.0	121.0
Straw	—	295.0
Leaves	—	260.0

[8] This is *Caragana arborescens*.

Table 4. Content of inhibitors (in RCU) in aqueous extracts at 1:10 from white campion (*Melandrium* [9] *album* (Mill.) Garcke.). pink (*Dianthus lanceolatus* Stev.) and black henbane (*Hyoscyamus niger* L.)

Sample	Campion	Pink	Henbane
Seeds	40.8	49.1	18.9
Wall of capsules (ovary)	85.5	141.5	121.0
Bladders	700.0	237.5	157.5
Leaves	—	—	198.0

One can assume that the inhibitors are formed in the walls of the pericarp, and have a genetic affinity with cholines in the leaves.

Accumulation of inhibitors in parts of the fruit occurs during their growth. After maturation of the fruit and the withering away of the pericarp, inhibitors are destroyed and gradually leached out by precipitation.

Table 5. Change in content of inhibitors (in RCU) in capsules of poppy (*Papaver somniferum* L.)

Sample	Capsules right After maturing	Capsules 30 days later
Walls of capsules	219.0	138.0
Seeds	9.4	12.0

For example, with the content of inhibitors in poppy capsules collected at differing phases of maturation, by 30 days it had decreased almost by half (Table 5).

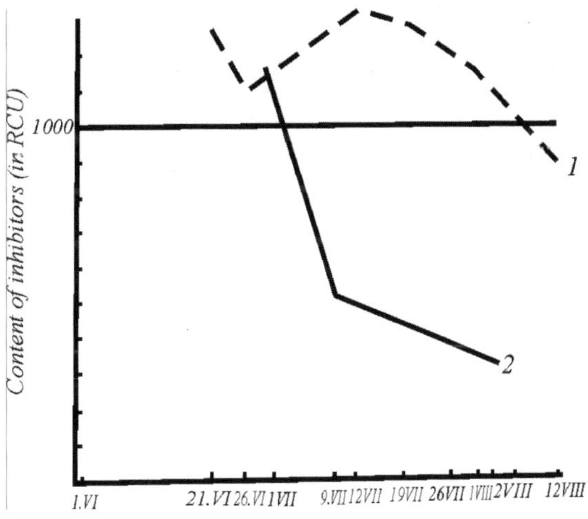

Figure 6. Dynamics of inhibitors in fruits of Tatarian bread (1) and Pontic colewort from southeastern Ukraine (2)

[9] Now regarded as *Silene latifolia*. As Grodzinskii's book was written 50 years ago, it is not surprising that there have been many changes to plant names, due either to taxonomic revision or priority ruling. Currently accepted names are footnoted. Obsolete names are termed "synonyms".

Figure 6 shows the dynamics of accumulation of inhibitors in the fruits of the two species of *Crambe* growing in southern Ukraine.

Although the amount of precipitation in the steppe after ripening of *Crambe* was very small, the content of water-soluble inhibitors in the fruit decreased by 1.5-3 times compared to the maximum.

Leaching of the inhibitors of seeds and fruits that remain on the plant can be quite complete. Table 6 shows the data regarding the content of inhibitors in the fruit of some plants collected immediately after maturation and after wintering in natural conditional (near Kiev).

Table 6. Leaching of inhibitors germination by precipitation in winter periods (content of inhibitors in RCU)

Plant species	In autumn	In spring
Box elder (*Acer negundo* L.)	439	23
Hoptree (*Ptelea trifoliata* L.)	570	16
Black locust (*Robinia pseudoacacia* L.)	≥1364*	13
False indigo-bush (*Amorpha fruticosa* L.)	≥1364*	22

* Note: 1364 RCU is the limit of sensitivity of the bioassay; germination is lacking.

Depending on the growth conditions (e.g. rainfall), on the biological characteristics of the plants (time of maturation and time of germination), as well as other circumstances, the accumulation of inhibitors may occur in different ways.

Table 7. Content of inhibitors in extracts from seeds of some anemochorous plants

Plant species	Content of inhibitors (in RCU)
Jack pine (*Pinus banksiana* Lamb.)	6.3
Scotch pine (*Pinus silvestris* L.)	10.5
Norway spruce (*Picea excelsa* Link.)	11.5 – 15.0
Chinese arborvitae (*Biota orientalis* Endl.)[10]	17.2
Wood anemone (*Anemone nemorosa* L.)	21.3 – 31.4
Canadian fleabane (*Erigeron canadensis* L.)	31.4
Lettuce (*Lactuca sativa* L.)	36.5
Siberian larch (*Larix sibirica* Led.)	59.0
Eastern salsify (*Tragopogon orientalis* L.)	53.7 – 91.0
Dandelion (*Taraxacum officinale* Wigg.)	83.0 – 128.0

So, for plants whose seeds are dispersed by wind and germinate soon after shedding, the small amount, or the absence of inhibitors is characteristic (Table 7).

[10] Now known as *Platycladus orientalis*

The seeds of these anemochorous plants, due to their wind dispersal, have low weight.

Therefore, even in possessing inhibitors, individually falling seeds or fruits cannot cause any serious allelopathic effect. Therefore, it is obvious that for seeds of anemochorous plants it is biologically inappropriate for them to accumulate inhibitors of germination and to release them to influence neighboring plants and their seedlings.

In anemochorous plants with a long interval between maturation and germination (maple, *Ptelea*, and others), inhibitors of germination accumulate in large quantities and are responsible for the delay in germination, which otherwise would be undesirable biologically from the standpoint of time.

The life form of plants that are tumbleweeds is sometimes referred to as anemochoric, and it is recognised as one of the devices for seed dispersal by wind, in that it rolls across the steppe with its spherical bushes with ripe seeds. Almost all of these types of plants are not eaten by animals due to their content of various harmful substances (Alekhin, 1939). Also note that the shape of tumbleweed develops in plants as a device for removing dead material of the aerial parts from the fixed point of growth, as these parts contain phytotoxic substances that cause self-suppression of plants. Tumbleweed plants are perennials with a taproot incapable of vegetative reproduction, so the accumulation of the dead above-ground parts, which are not eaten by animals, can cause self-suppression. The dissemination of seeds from the tumbleweed bushes occurs differentially - they either shed seed while still in place when the wind shakes the bush, and has not had time to break away from the root crown, or when not shed, they are blown by the wind to places inappropriate for germination.. For example, in Khomutovskaya Steppe, it was noted that the brush of Tatarian bread, richly strewn with seeds, spread typically into ravines, river valleys, bushes, where *Crambe* never grows.

One could assume that the fruits and seeds of plants belonging to the life form of the tumbleweed will contain a large quantity of inhibitors (Table 8).

Table. 8. Content of inhibitors in fruits and seeds of tumbleweed plants

Plant species	Content of inhibitors (in RCU)
Trinia henningii[11]	253-263
Trinia hispida Hoffm.	246
Tatarian bread (*Crambe tataria* Sebeök)	51-1100
Serratula xeranthemoides[12] M.B.	411
Blue eryngo (*Eryngium planum* L.)	422
Baby's breath (*Gypsophila paniculata* L.)	470
Statice (*Goniolimon tataricum* (L.) Boiss.)	272-490
Sea lavender (*Limonium latifolium* (Sm.) Ktze.)	525

[11] Now known as *T. multicaulis*

[12] Now known as *Klasea erucifolia*

The biological purpose of inhibitors in the fruit of tumbleweed is, apparently, besides for keeping the seeds in dormancy, also to inhibit the growth of surrounding vegetation. As a rule, the fruits of tumbleweed fall far from the mother plant and fall into thick swards with the litter of other steppe plants. Under such conditions, the successful germination and plant establishment are possible where the fruit provides clearing of a living space by the action of releasing its inhibitors. It is important that fruitlets are not shed one by one, but in groups, as so often happens in *Crambe*, as it thus creates a small centre of concentrated introduction of a new component into the cenosis.

Projectile plants (Levina, 1957) sow their seeds, from goblet-shaped capsules opening on top, due to vibrations caused by wind gusts, and which contain in the walls of these capsules very large quantities of inhibitors; thus from the start seeds are almost "untainted" and capable of sprouting (Table. 9).

Table 9. Content of germination inhibitors in seeds and dry pericarps (ovary + calyx) of some projectile plants

Plant species	Content of inhibitors (in RCU)	
	In seeds	In pericarp
Opium poppy (*Papaver somniferum* L.)	23.5	405
Corn poppy (*P. rhoeus* L.)	14.5 – 26.2	930
Clustered bellflower (*Campanula glomerata* L.)	32.5	988
(*Otites chersonensis* (Zapal.) Klok.)[13]	9.2 – 34.1	128 – 475
Carnation (*Dianthus caryophyllus* L.)	10.2 – 51.0	355
Lanceolate carnation (*D. lanceolatus* Stev.)	24.0 – 68.0	396
Long-scaled pink (*D. pseudoarmeria* M.B.)	63.0	384
Broadleaved campion (*Silene latifolia* (Mill.) Rendl. *et* Britt.)	20.5	63
Orange mullein (*Verbascum phlomoides* L.)	34.0	1250
Purple mullein (*V. pheoniceum* L.)	17.9	240
White mullein (*V. lychnitis* L) right after ripening	34.0	1255
As above (in two months)	255	260

Seeds of *Silene*[14], campions and others extracted right after maturing of the bolls were able to germinate immediately, albeit in small quantities (15%). Washing of these seeds does not significantly change their germination.

Plant seeds under natural conditions, after precipitation, gradually absorb water that has extracted inhibitors from the walls of the capsules; they become dormant and do not germinate after being removed from the capsule. This occurs already by late autumn.

---

[13] Now known as *Silene exaltata*

[14] The name given by Grodzinskii is *ушанка*, a local name, which translates as earflap or earlobe, in reference to the thus shaped seeds of *Silene otites*

Thus, due to the presence of water-soluble, mobile inhibitors in the walls of the capsule, the plant sows seeds of differing qualities and biological properties. The first batch of seeds that are spilled in the summer, just after the boll maturation and opening of corresponding apertures, do not contain inhibitors, and germinate immediately if there are the right conditions. Young plants have time to develop before winter and to survive the winter hibernation satisfactorily. The batches of seeds, sown in late autumn, which having absorbed a large quantities of inhibitors due to autumn rains, do not germinate, even if there are favorable conditions, and thus. In this case, the seedlings do not have time to develop and gain a foothold for overwintering. Such action of inhibitors is biologically beneficial to plants and could have developed and consolidated evolutionarily. Of course, this is not the only mechanism for allowing resting and creating seeds of heterogeneous quality.

Projectile plants usually have very small seeds, and disseminate them nearby - a maximum of a few decimeters from the mother plant. Therefore, the amount of inhibitors spread with them is very small and probably does not play any allelopathic role. Projectile seeds fall under the allelopathic protection of the litter and excretions of their own mother plant, and thus inhibitors in the seeds are not required for chemical interaction with other plants.

Some plants of the legume family spread their seeds as a result of tightening the valves of the pod, for example Siberian peashrub and Russian peashrub, which have almost the same content of water-soluble inhibitors in seeds and pericarp (Table 10).

Table 10. Content of inhibitors in fruits of legumes (in RCU)

Plant species	Seeds	Walls of legume
Siberian peashrub (*Caragana arborescens* Lam.)	80	86
Russian peashrub (*Caragana frutex* C. Koch.)	212	246

It is interesting to note that Russian peashrub has three times the amount of inhibitors of Siberian peashrub, which has been quite intensively studied (Blagoveshchenskii, 1953, Kolobkova, Kudryashov, 1956, 1957[15]).

Other legumes, in contrast, are characterised by very large differences in the content of water-soluble inhibitors in the seeds and the fruit shells (Table 11).

Table 11. Content of inhibitors in fruits of legumes

Plant species	Content of inhibitors (in RCU)	
	In seeds	In walls of legume
Field pea (*Pisum sativum* L.)	31.9	425
Sun-dial lupin (*Lupinus perennis* L.)	9.5	209
Spanish broom (*Spartium junceum* L.)	4.9	109
Horse bean (*Vicia faba* L.)	24.5	245

[15] 1957 reference not in bibliography.

Evidently, the explanation for this phenomenon can be similar to that for cruciferous or projectile plants, i.e., the high content of inhibitors in fruit shells serves to delay seed germination until opening of the fruit.

The family of legumes has such a variety of adaptations for seed distribution - from the underground fruit of the peanut, explosive fruits of peashrubs, air-filled bladders (*Anthyllis* L.) to various fruits that mature near the parent (just fall from the plant), that the biological role of inhibitors, in particular their allelopathic role, needs to be discussed in each case.

One might think, for example, that the value of allelopathic inhibitory substances in the wings of the indehiscent pods of Don sainfoin (*Onobrychis tanaititsa* Spreng.)[16] at 800 RCU[17], sainfoin (*O. viciifolia* Scop.) 424 RCU, indigo bush (*Amorpha fruticosa* L.) ≥1364 RCU, etc., is in creating, as in the case of life form of tumbleweeds, a specific biochemical sphere around the germinating seed.

In the case of projectile dehiscent legumes, the valves themselves eventually fall into the overall plant litter, and enhance the biological activity of the litter.

Seeds of some plants, that have seeds spread actively, such as various kinds of touch-me-not (*Impatiens* L.), do not usually have inhibitors of germination, while the walls of succulent fruit, which can release the seeds due to a sudden change in turgour pressure, contain a lot of them. For example, the ornamental plant *Impatiens balsamina* L. contains in the seeds 5.22 RCU, but in the dry fruit ≥1364 RCU, i.e., beyond the sensitivity of the bioassay.

Plants, whose seeds in wet weather become gelatinous, stick to the ground or to an animal, and are thus dispersed (e.g., flax species, sage), and excrete in aqueous solution relatively small amount of inhibitors (Table 12). The calyx, within which are located capsules, pods, etc., in contrast, contain high amounts of inhibitors.

As stated, the mucilage formed by these seeds, in most cases, causes almost no delay in the germination of seeds of radish and other plants. The slimy material consists mainly of polysaccharides, themselves excellent substrate for the development of many microorganisms, but, nonetheless, it remains perfectly unaffected for weeks in dishes, where the seeds have germinated without going off or being covered with mould. Obviously, this mucilage contains strong antibiotic substances inactive with respect to higher plants. Known therapeutic effects of decoctions of flax seed may be due to the antimicrobial properties of mucilage. This should attract the attention of researchers engaged in the search for antimicrobial substances.

The mucilage produced by pectic substances, also does not significantly delay the germination of seeds. Thus, the inhibitory activity of extracts from the seeds of quince (*Cydonia oblonga* Mill.) corresponded to only 31 RCU.

Fleshy, juicy, sweet fruits, that serve as an attractant for birds and mammals that eat them and are transferred to their gut, have hard, indigestible seeds surrounded by

---

[16] Now regarded as *O. arenaria* ssp. *arenaria* (Kit.) DC.

[17] The RCU (relative coumarin unit) is discussed in the Appendix.

a bony endocarp, and almost without exception, it contains high amounts of inhibitory substances (Table 13).

Table 12. Content of inhibitors in gelatinous seeds

	Content in RCU	
	In seeds	In fruit cover
Austrian flax (*Linum austriacum* L.)	10-20	47-255
Field flax (*L. usitatissimum* L.)	13.9	122.5
Austrian sage (*Salvia austriaca* L.)	21.5	200.0
Steppe sage (*Salvia tesquicola* Klok. *et* Pobed.)	37.5	59.0
Nodding sage (*Salvia nutans* L.)	240	345
Meadow sage (*Salvia pratensis* L.)	92	—
Broadleaf plantain (*Plantago major* L.)	17.2	110
Hoary plantain (*P. stepposa* Kupr.)[18]	17.2	746
Pale madwort (*Psilonema calycinum* (L.) C.A.M.)[19]	12.0	48.7
Hoary cress (*Cardaria draba* (L.) Desr.)	63.5 – 137.8	1050
Clasping pepperweed (*Lepidium perfoliatum* L.)	72.3	95

Table 13, in addition to the data obtained in bioassay with radish, gives the magnitude of the growth of cut discs of cotyledons of sunflower. We have used this test, because in extracts from fresh and dried fruits of this kind, containing many types of sugar (fructose, glucose, sucrose, etc.), which when applied at the concentrations being used, may retard the germination of radish, whereas with the growth of discs, the sugar has no particular effect. This makes it possible to detect inhibitors of a non-carbohydrate nature.

The fate of the inhibitors in the body of the animal is unknown. We can assume that they are largely destroyed, as animal excretions themselves have phytotoxic properties. Seeds in faeces after breakdown, washing with rain, etc., acquire the ability to germinate. Many of these plants do not germinate until the following year, and artificial seeding may require stratification. Active substances in excrement (telergony) can perform some allelopathic activities of creating space for seed germination. Probably this method of colonisation is of great importance in the chemical interaction of plants.

As for the inhibitors in the fleshy fruit that are not eaten by animals, they are shed en masse from the branches of trees or shrubs, and cover the soil; they can have a direct chemical effect on seeds, seedlings and adult plants of alien species that have invaded under the canopy of the donor plant.

Anyway, whatever the fate of ripe juicy fruits, they contain inhibitors that are of great allelopathic importance.

---

[18] Now known as *P. media*

[19] Now known as *Alyssum alyssoides*

Table 13. Content of inhibitors in extracts from fruits of berry, stone fruit and orchard stock

Plant name	Content of inhibitors in RCU from extracts 1:10 (from dry material)	Growth of segments from sunflower cotyledons (as % of control)
Black currant	26.6-142	101.5
Stone-berry	473	—
Apple	539	98.0
Mahaleb cherry	594-990	—
Raspberry	613-635	—
Grape	645	65.3
Red elderberry	800	75.2
Apricot	890	5.0
Plum	890	73.0
Pear	980	66.6
Wild strawberry	1100	—
Black mulberry	1240	10.0

In contrast to these kinds of plants, are those whose seeds are spread by ants (myrmecochory), such as the species of adonis (*Adonis* L.), which contain a lot of inhibitors in the seeds but less in the fleshy nutritious pericarp. So in extracts from Volga adonis (*Adonis wolgensis* Stev.) the extracts contained in RCU: from whole fruits – 322, from purified seeds – 890, from the pericarp - 74.

Perhaps this anomaly is not connected to the dispersal process, but to characteristics of some members of the buttercup family where the inhibitors decrease in the concentration from the interior of the fruit to the outside. For example, in oriental virginsbower (*Clematis orientalis* L.) the seeds registered 65.2 RCU, and the shells of the fruit - 17.4 RCU; in golden clematis (*C. tangutica* Korsch.) - 209 and 63.6 RCU respectively.

There are especially a lot of inhibitors in all parts of the fruits of the cucurbits (Table 14).

From the literature, it is known that germination inhibitors in the fruits of pumpkin are important for creating conditions of forced dormancy before the destruction of the fruit (Molotkovskii, 1949). However, the role of inhibitors is not restricted to this. Seeping from rotting fruit in the soil, inhibiting substances destroy or inhibit the plants and seeds in the place where the fruit is, and thereby prepare a place to establish their own seed, which is due to the adaptation that it is less sensitive to these inhibitors. It is possible that at concentrations harmful to other plants, there is stimulation of germination of seed of the donor. Thus, we have analysed part of the fruit of zucchini (*Cucurbita pepo* var. *turcica* Zhiteneva) within which in spring its own seeds germinated, but it still contained ≥1364 relative units, i.e., it completely inhibited the germination of radish.

Table. 14. Content of inhibitors in aqueous extracts from parts of fruits in the gourd family (Cucurbitaceae L.)

Plant species	Content of inhibitors in RCU		
	In seeds	In flesh	In pericarp
Summer squash (*Cucurbita pepo* L.)	17.4-44.0	≥1364	≥1364
Cucumber (*Cucumis sativus* L.)	87.0	–	≥1364
Watermelon (*Citrullus vulgaris* Schrad.)	85.0	1290	≥1364

The inhibitors from the pumpkin family do not belong to the sugars, amino acids, tannins and nucleotides.

Such a mechanism of using inhibitors of germination in the struggle for existence is inherent in many other fruits, fleshy and dry. For example, in the case of the falling of the fruits to the ground (basicarpy)[20] creates a more or less dense layer of material with dead leaves and other litter from which inhibitory substances are freely extracted with leaching. From experience, it is known that if a lot of beet seed is left for a long time in the rain, the grass will not grow in this place for a few years. The beet also has the well known property of being multi-seeded and there is a positive impact on the quality of the crop with large seed balls, although seed-ball tissue - woody pericarp - does not contain nutrients for seed substances (Maksimovich, 1957). Such an accumulation of pericarp material with a high content of inhibitory substances can be explained by the protective role of this substance in the process of seed germination and plant colonisation in the phytocoenosis (Isip, 1939). Inhibition of seed germination in cockle (*Agrostemma* L.) on filter paper under the influence of aggregates of fruits from beet was noted by M. Evenari (1949), who believed that the absence of corn cockle in beet fields can be explained by the inhibitory effect of excretions from the seeds. However, G. Grümmer (1957) questions this view, as the inhibitory effect on the germination of cockle seeds disappears when soil is present.

It is emphasised in the literature that for many other plants, they have the property of group sprouting when a bunch of seeds or fruits are sown (spruce, pine, cedar).

The positive role of denser sowing or planting is known from the practice of forestry, and is thus explained by the cooperative allelopathic action of individuals. Later, after the plants reach a certain growth stage, in many instances the joint growth of many specimens in a small space becomes harmful for them, which leads to the death of some and to their thinning.

The huge surplus of seeds that die without germinating, and especially the great bulk of organic material of their fruits, at first glance seem to be unnecessary to the plant and, therefore, contrary to the principle of "economy of nature", but can be explained somewhat by a need to develop allelopathic potential, particularly for "its own" environment, conducive to its own offspring and harmful for other plants. Such

---

[20] The term "basicarpy' refers to the development of flowers and fruits above ground, but the terminology in this area is complex and often confused. A better term here might be "barochory", which refers to the dispersal of propagules by gravity alone.

conditions are found in nature often enough that it may be consolidated phylogenetically.

In a single seed, the inhibitors are too weak for its mass to conquer living space, but the combined effect of many seeds inhibitors, in combination with inhibitory substance from residues of fruits, leaves, roots, etc., is an active factor of considerable strength.

Thus, the life of the plant begins with the fact that the seed or the remains of the pericarp as formed by the mother plant, release into the environment physiologically active substances, toxic or beneficial to other plants. Over time, these substances are supported later by seedling root exudates and other types of excretion, that create a special chemical sphere around the plant. The strength and impermeability of the sphere increases when the same plants form a clump, family, patch, microassociation, group. Thus, germination inhibitors are "the first wave of attack".

For allelopathic effects, it is important that the inhibitory substances are persistent, are not absorbed completely by the soil and retain their activity for quite a long time. Presumably, this occurs for most of the inhibitors of germination of seeds and fruits. For example, the germination inhibitors from the fruit of *Crambe tataria* under aerobic composting conditions (fruitlets maintained at 50% humidity at room temperature) retain activity for six months. They are slightly absorbed by the soil taken from the natural sites growing *Crambe* (Grodzinskii, Kuznetsova, Musatenko, 1960). Under anaerobic conditions, inhibitors completely lose their activity in 12-15 days. They are very strongly absorbed by soils that are different from the carbonate-rich chernozem, which is common with *Crambe*.

Inhibitors in the fruits and seeds of cultivated plants can be of considerable practical importance. It has been shown that to improve the germination of many forest plants they should be washed with water (Misnik, 1949).

Soaking the seeds of many vegetable crops before planting also has a positive effect on emergence. Seeds which respond especially well are those where soaking decreases the content of endogenous inhibitors or washes inhibitors from the pericarp - beets, watermelon, melon, cucumber, pumpkin (Kasimov, 1959)[21]. Large seeds of legumes (beans, peas, broad beans) with soaking have reduced germination; results with cereal grasses remain undetermined. Obviously, the removal of inhibitors should proceed only to a certain level, where their content in the seeds starts to exert a stimulating effect.

Adjusting the level of endogenous inhibitors in seed stock will in the future have practical application on a larger scale than at present, and in some cases (e.g. in pre-winter sowing) it is advisable to enrich the seed with germination inhibitors to prevent the appearance of shoots at the wrong time.

Seed pre-treatment with bacterial preparations, such as azotobacterin, nitragin, phosphobacterin, etc., should take into account the fact that the seeds and fruits can often contain antibacterial agents that reduce or completely suppress the proliferation of strains used for infection.

---

[21] There is no reference for this citation, and one could not be located.

Bowen (1961) showed that the excretions of clover seeds are very toxic to *Rhizobium trifolii* and a number of other microorganisms. The same applies to the action of seeds of other legumes. This effect is particularly enhanced at the time of seed germination.

Root exudates and their role in allelopathy

The problem of root exudates has occupied many researchers. Mostly studied are substances released into the aqueous solution. It is believed that secretions into an aqueous nutrient medium and the soil are of the same character, so the method of water cultures of plants, especially in a sterile environment, is suitable for the study of this matter (Chalvignak, 1958).

There are ideas about active exudations of solutions from roots. Thus F. Eberhardt (1955) observed in a humid chamber the appearance of drops containing scopoletin on the root hairs of oats. Other studies have refuted the possibility of active excretion of substances from living, intact roots. Criticism of the experiments of F. Eberhardt was made by Martin (1956), who believed that the excretion of substances is due to the action of certain unfavorable conditions, when the cells of the root are unable to "retain" matter, and it is free to diffuse into the environment. The droplets that were observed by Eberhardt are condensed moisture and can also form on the fibres of cotton wool. Furthermore, Martin (1963) argues that the loss of substances from the root is associated with impaired metabolism and is not a normal expression of cell activity. Various factors that hinder the supply of energy to the cells, required for the cells to maintain their content, increase the discharge of substances. The author has determined by more sensitive process that the excretion of scopoletin by oat roots occurs according to changes in the external environment. The excretion of scopoletin is increased by insufficient aeration of the medium, when maintaining the roots in pure water or single salt solution at supraoptimal temperature, and by the addition of filtrates from cultures of certain microorganisms and substances of microbial origin. The experiments indicate that the excretion of scopoletin and apparently other substances is due to a breach of the cell metabolism, even if the effects of these factors do not seem to impact on the plant.

Through the actions of dinitrophenol sodium azide and cyanide, P, Martin (1963) showed an increase in the loss of scopoletin by oat shoots, and the excretion of labeled sucrose, fructose and unidentified substances from soybean roots, even at low concentrations (10^{-7} M) of these toxins. In this regard, it is believed that with the uncoupling of oxidative phosphorylation, ATP levels decrease, leading to a decrease in the retention capacity of the cells. Any loss occurs due to the increased permeability of the protoplast. The author, however, was at a loss to explain why the composition of root exudates of soybean extracts proved to be non-identical with the composition of the roots. In addition, a substances was found in the excretions that was absent in the roots. H. Börner (1956 a, b) also discusses the absence of root exudates in plants. In studies with rye, wheat and barley, he could not find root exudates, if the plants were under normal physiological conditions. Some authors allow the discharge of substances as due only to the death of the cells of root hairs, and epidermis and root cap.

Yet, most researchers tend to believe that the excretion of substances by roots goes on as an active process, as opposed to the process of absorption. Namely, organic excretions appear coated as a thin layer on the root surface, and can be exchanged for the ions in solution or adsorbed to soil colloids (Gellerman, 1950). At present, most contemporary theories of uptake of substances envisage the exchange of absorbed molecules or ions with those excreted by the root - carbonate ions, bicarbonate, hydrogen molecules and various organic compounds.

B. Frenzel (1957, 1960) considers the excretion of amino acids and amides by roots of the sunflower as a metabolic activity, a normal function of living cells, and "diffusion plays only a subordinate role".

Reasons for differences in views of the individual authors are, in our opinion, the differences in the methods used. For root exudates, they should be isolated from the action of microorganisms and resorption by the plant donor. This can be achieved with sterile cultures, frequent changes of solutions or the use of adsorbents in solutions, and the use labelled atoms.

We believe that the existence of root exudates is not in doubt. Their presence is perfectly illustrated by experiments using tracer methods.

With regard to the excretion mechanism, it is likely varied. Excretion should include losses due to ordinary diffusion, and the withering away and shedding of some covering tissues and finally the normal physiological release of active substances in the environment. In order for a plant to make contact with the soil, it is forced to develop a huge root surface area that is poorly protected from external influences. To be able to absorb substances, the protoplasm of roots must be exposed, but it also leads to the fact that it is easy to lose substances and water.

Plant roots are usually electronegatively charged relative to the soil, so they absorb and retain cations more easily than anions. Anions, in a simple way, are "extricated" from the root by electrostatic forces; they are consumed in large quantities for the exchange of the essential plant mineral anions (phosphate, sulphates, nitrates, etc.) from the soil, which is why the root secretions of plants are dominated by substances of an acidic nature.

Various adverse effects that disrupt the normal balance of metabolic processes, and increase the loss of substances from roots. Generally it can be said that the more that the conditions deviate from the norm, the more substances are released. For example, soil drying prior to wilting of plants and subsequent watering promote enhanced release of amino acids and reduced compounds (Katznelson *et al.*, 1955), and anaerobic conditions increase the excretion of sugars and organic acids, which are intermediates of the Krebs cycle (Grineva, 1961). In contrast, B. Frenzel (1957) found that the maximum release of amino acids and amides from sunflower roots in sterile conditions occurs under the optimal combination of growth conditions for sunflower.

There is really no contradiction between these data. The total amount of excreted substances will be more in normally growing plants, whereas the intensity of excretion in times of adverse effects will be greater with specimens that are suppressed.

The majority of families of higher plants have representatives for which the release of phytotoxic substances from roots is known or suspected. The well-known excretion of organic acids has been found in the last century in connection with the study of the so-called solvent power of roots. As is known, many organic acids are toxic to plants, or in any case have a quantity of physiologically active substances. Therefore, organic acids secreted by roots can be of direct phytocenotic value. Among the organic acids excreted by various plants that have been identified are oxalic acid, succinic acid, fumaric acid and other acids of the Krebs cycle (Grineva, 1961; Vančura, Hovadik 1963), trans-cinnamic acid (Bonner, Galston, 1944), chlorogenic acid (Politis, 1948) and others.

Organic acids not only change the pH of the environment and increase the solubility of minerals, but also have an impact on the life of the soil microflora.

The fact that root secretions of plants also contain nitrogen compounds has long been known (Virtanen, Hausen, 1931; Virtanen, Laine, 1935; Virtanen, 1947). Using methods of sterile culture and paper chromatographic separation of substances, it has been possible to examine the quantity and qualitative composition of amino acids and amides excreted by different plants (Kandler, 1951; Katznelson *et al.*, 1955; Linskens, Knapp, 1955; Parkinson, 1955; Andal *et al.*, 1956; Frenzel, 1959; Carré, 1962).

Among those identified are glutamic, aminobutyric, and aspartic acids, amides of the latter, leucine, valine, alanine, serine, tryptophan, cysteine, glycine, lysine, phenylalanine, proline, threonine, homoserine. Methionine and other sulfur-containing amino acids are found in small quantities. The quantity of excreted amino acids depends on their concentration in the roots themselves, on nitrogen nutrition and aeration. Amine and imine nitrogen excretion represents about 70% of the total nitrogen excreted. The amount of amino acids secreted for 3-4 weeks may contain 0.07-1.03 mg per gram of root. The older the plant, the more that amino acids predominate in the excretions (Rovira, 1956). G. Frank (1954) found that at the end of the growing season plants lose huge amounts of nitrogen (e.g. up to 30% of the maximum accumulated nitrogen in *Arabidopsis thaliana* (L.) Heynh. and 24% in *Oenothera suaveolens* Desf.). This is explained by leaching of amino acids from the roots, and possibly also by the release of nitrogen in gas form from the leaves.

In the root secretions, there are also found other nitrogenous compounds: ammonia (Pryanishnikov, 1951), nucleotides (Lundegårdh, Stenlid, 1944) and some derivatives of nucleic acids (Fries, Forsman, 1951; Anderson, 1957). Excretion of nitrogenous bases reaches 5 mg per 1 g of the dry weight of the roots. Among amino sugars are found glucosamine, galactosamine and two poorly understood amino sugars (Vančura, Hovadik, 1963), as well as traces of the urea.

A number of studies have shown the capacity of the roots of young plants to release sugars into the environment (Patrick; 1955, Linskens, Knapp, 1955; Rovira, 1956; Katznelson *et al.*, 1955; Martin, 1956, 1957), which, of course, are quickly processed by microorganisms and affect the growth of neighboring plants only indirectly. It seems that the roots of older plants are unable to excrete sugar.

Vančura, Hovadik (1963) found that in the root secretions of barley and wheat the sugars were mostly represented by oligosaccharides, galactose, glucose, arabinose, xylose and rhamnose; in cucumber by fructose and arabinose; in kohlrabi and peppers by fructose; and in swede by xylose. In the excretions of barley they found galacturonic acid, and in wheat, galacturonic and glucuronic acids.

Of particular importance to the mutual influence of higher plants is the excretion by roots of various biologically active substances, notably enzymes, vitamins, alkaloids, glycosides, antibiotics, natural and synthetic growth substances, etc.

Research by E.F. Kuprevich (1949, 1951), N. A. Krasil'nikov (1952) established that roots lose by exosmosis various plants enzymes, including catalase, tyrosinase, phenolase, asparaginase, urease, amylase, invertase, cellulase, protease, lipase. W. Flaig and R. Kaul (1956), E.I. Ratner, SA Samoilova (1954), and I.I. Kolosov (1962) established the excretion of a number of enzymes that regulate phosphorus nutrition of plants.

It is common in recent years to determine the enzymatic activity of soil (often called "biological activity of the soil"), which is partially attributed to the presence of enzymes from plant root exudates (Peive, 1961). As suggested by N.V. Peterson (1961), enzymes that are secreted directly by roots of higher plants are weaker than those produced by rhizosphere microflora. Whatever the case, the presence of enzymes in the environment of the root nutrition of plants is of great allelopathic importance.

The excretion of vitamins is essential for the development of the rhizosphere microflora and mycorrhizae, and hence for the joint plant growth and the creation of associations (Ovcharov, 1961). Studies by West 1939) found the excretion of thiamine (B_1) and biotin in roots of young plants of flax. On this same topic, M.A. Chalvignak 1962) established the excretion of folic, nicotinic and pantothenic acids, riboflavin (B_2), and found no excretion of pyridoxine and vitamin B_{12}. Other researchers have shown for other plants, there can be excretion of pyridoxine (B_6), inositol, niacin (B_3), pantothenic acid (B_5), and riboflavin (Meshkov, 1952; Winter, Rümker, 1952; Harley, 1952). Secretion of vitamins by roots depends on the species and the age of the plant. Thus, in the experiments of N.A. Krasil'nikov (1958), peas excrete more thiamin and biotin than wheat or corn; 10-day seedlings of pea excreted several times more than did 45 day-old plants. Thiamin, riboflavin, nicotinic acid, carotene have been discovered in soils, and their concentration is determined primarily by microbial activity, however, it depends also on the species and state of the plants. So, N. E. Gvamichava (1963) showed that, in a humus-sulphate soil of Samgori, B vitamins were found in larger amounts under vines than under the grass. The soil under chlorotic plants contained less thiamine, riboflavin and pyridoxine and more nicotinic acid than from soil of the rhizosphere under healthy vines; pantothenic acid content was almost the same.

Of considerable allelopathic importance can be glucosides, which are sometimes found in the root secretions. E.N. Mishustin and A.N. Naumov (1955) found that the roots of alfalfa release into the soil a significant amount of saponins, which eventually leads to thinning of the alfalfa sward growth and reduces subsequent crops (cotton). Some well-known excreted lactones of acids and their glycosides are, e.g. coumarin and scopoletin (7-hydroxy-6-methoxycoumarin,

Eberhardt, 1954, 1955; Martin, 1956. 1957, 1958, 1963), which are active growth inhibitors.

Dying remnants of the root system of peach endow the soil with amygdalin and its degradation products - hydrocyanic acid and benzaldehyde (Börner, 1960 a, b), which prevent the regrowth of plantations of these trees.

Alkaloids have been found in the root secretions by K. Mothes (1955). In the root secretions of encelia (*Encelia farinosa* A. Gray) were discovered 3-acetyl-6-methoxybenzaldehyde (Bonner, 1950), in flax - hydrocyanic acid (Timonin, 1941), in couch grass - agropyrene (Laštuvka, 1955), in apple - phlorizin and its derivatives, in lovegrass - catechol (Scheffer *et al.*, 1962). An important allelopathic role is attributed to all of these substances (Woods, 1960).

The appearance of flavanones, that are known to be very active, was observed in the medium growing isolated roots in experiments with wheat and pea (Lundegårdh, Stenlid, 1944) and peanuts (Petrii, Chrasti 1955); however, the flavanones may have been released through the upper cut parts of the roots.

There are many publications on the excretion of toxic plant substances, of which the chemical nature is unknown. For example, H. Börner and B. Rademacher (1957 a,b) and H. Börner (1958 a), using an original method of flow culture, established the excretion by whole intact roots of flax, oats and other plants, phytotoxic substances which retard the growth of flax or wild mustard. These substances are apparently not phenolics, as phenolics were not detected previously in concentrated root exudate solutions. However, the nature of these compounds has not been studied yet.

Physiological effects of root exudates are primarily due to metabolic changes in the roots of the acceptor plant, in the acceleration or deceleration of flow of water and nutrients. As they penetrate into the plant, physiologically active excretions have diverse effects on metabolism, which are expressed in general in promoting or slowing the growth processes.

On the other hand, root exudates have large indirect effects on the physiology of plants through changes to the soil microflora. Numerous soil inhabitants live mainly as a result of root exudates and dead organic matter in the soil, so the qualitative composition and quantity of root exudates play a key role in determining the composition, strength and activity of microflora. The role of rhizosphere microflora in the life of higher plants cannot be overestimated.

V. Vančura and A. Hovadik (1963) in experiments with seedlings in sterile and semi-sterile water cultures with six species showed that the quantity of root exudates corresponds to 7-10% of the dry weight of the aerial parts of the plants. In barley and wheat it amounted to 400-500 μg per single seedling. The dry residue of barley root exudates contained 19.1% ash, 9.1% reducing agents, 0.2% of volatile acids, 17.2% residual acids and 1% of total nitrogen.

Not all organic matter in the root exudates is active physiologically; much of it is present as carbohydrates (fiber) and other compounds that are inactive, or are rather quickly destroyed by the soil microflora. However, as we have seen above, a lot of physiologically active substances in the root secretions of found from a wide variety of plants have been identified, and there are even more substances that are unidentified. Active substances are described from quite credible cases of allelopathy with roots, so the search for cholines in these types of secretions is of great importance for establishing the extent and the role of allelopathy in the plant world.

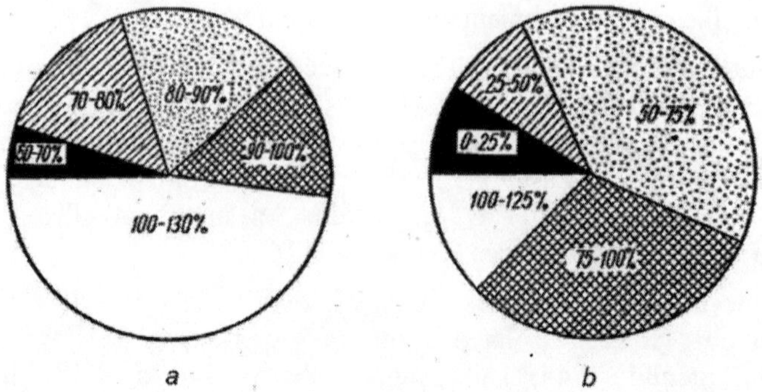

Figure 7. Prevalence of cholines in volatile (a) and water-soluble (b) root excretions from 148 species of herbaceous plants (indicated by germination of radish in bioassay with percentage for K-50%)

In a study of root exudates of many plant species, obtained from young and old individuals, we have found a quite wide range of phytotoxic substances. The distribution of the species according to the activity of root exudates is shown in Figure 7.

Despite the relatively large number of studied plants - about 150 species - we have not found any connection between toxicity of the root exudates and the taxonomic position of the plants or their environment. Further research is needed.

Bioassays determined that excretions of particular toxicity were from root systems of awnless brome grass (*Zerna inermis* (Leys.) Lindm.[22]), couch grass (*Elytrigia repens* (L.) P.B.). peas, yellow and blue lupine, in root exudates of bushgrass (*Calamagrostis epigeios* (L.) Roth.), hairy feathergrass (*Stipa capillata* L.), creeping thistle (*Cirsium arvensis* (L.) Scop.), oats, wheat, rye and barley.

Root exudates of maize were almost harmless. Root exudates of field bindweed (*Convolvulus arvensis* L.), Canadian flea-bane (*Erigeron canadensis* L.) and odorless chamomile (*Chamaemellum inodorum* (L.) Vis.) had no effect on the growth of segments of rhizomes of couch grass.

Root exudates were quite resistant to heat and decomposition by soil microflora. Thus, in experiments of T. N. Filippovich, non-sterile soil was added as a primer to the solution of root exudates of brome or couch grass, and after only two or three days at room temperature there followed a significant degradation of toxic substances. In studying the accumulation of cholines in aqueous culture under roots of couch grass and other plants, it has also been found that the maximal accumulation falls within the second to fourth day, after which the solution was observed to

---

[22] Now known as *Bromus inermis* Leyss.

decrease in activity due to the destruction of toxic substances by microflora that developed.

Extracts from whole intact roots gently dug out of the soil are often toxic in bioassay, but to a lesser degree than the same extract (the leachate) from whole leaf. Phytoncides, i.e. volatile vapours from the pounded roots often stimulate seed germination. To illustrate these proposals, Table 15 shows the data determined for these substances at the same time for the same plants.

The activity of soluble root exudates (leachates), as can be seen from Table 15, is slightly inversely correlated with the activity of their phytoncides: the more inhibitory water-soluble cholines (from the top of the table), the stronger the stimulatory effect of phytoncides[23]. There is almost no correlation between the activity of root exudates and secretions from the leaves.

Table 15. Activity of water-soluble excretions of roots and leaves and volatile phytoncides of roots of steppe plants

Species of plant	Water-soluble substances as RCU		Phytoncides of roots	
	Roots	Leaves	Radish at K-50%	RPU*
Yellow reseda (*Reseda lutea* L.)	720	850	110.0	-0.09
Couch grass (*Elytrigia repens* (L.) P.B.	448	1330	146.0	-0.32
Jurinea (*Jurinea dubia* (Korsch.) Iljin.[24]	330	1350	133.0	-0.25
Volga adonis (*Adonis wolgensis* Stev.)	310	1290	115.8	-0.14
Narrow-leaf meadow-grass (*Poa angustifolia* L.)	218	59	86.0	+0.16
Goldilocks (*Linosyris dracunculoides*)[25]	168	480	102.7	-0.03
Bunchgrass (*Festuca sulcata* Hack.)[26]	157	226	113.0	-0.11
Awnless brome grass (*Zerna riparia* Rehm.)[27]	118	180	61.0	+0.64
Iris (*Iris halophila* Pall.)	115	25	99.0	+0.01
Fernleaf peony (*Paeonia tenuifolia* L.)	108	—	74.0	+0.35
Siberian bellflower (*Campanula sibirica* L.)	98	1310	108.0	-0.08
Tatarian bread (*Crambe tataria* Sebeök.)	77	—	116.0	-0.14
Iris (*Iris taurica* Lodd.)	65	61	96.5	+0.04
Hair-like feather grass (*Stipa capillata* L.)	54	246	115.0	-0.13
Pontic colewort (*Crambe pontica* Stev.)	30.6	542	76.0	+0.32

* Note RPU - relative phytoncide units.

[23] Note that with the RCU scale, a higher value indicates increasing toxicity; and with the RPU scale a negative value indicates stimulation

[24] Now known as *Jurinea consanguinea* ssp. *arachnoidea*

[25] Now known as *Aster linosyris* ssp. *dracunculoides*

[26] Now known as *Festuca rupicola*

[27] Now known as *Bromus inermis*

Calculated using the formula: $RPU = \frac{100}{\textit{Germination of radish at K50\% (in \%)}} - 1$

Generally, as a rule, all types of excretion - water-soluble and volatile, vital and posthumous - are subject to the same laws: they are most active in the leaves and least active in the roots; stems occupy an intermediate position. This pattern deviates only in very rare cases, which are therefore of particular interest for study. In this connection, there is the idea that other types of secretions may be more important in respect of allelopathy, as with root exudates, as has been sometimes represented in the literature, for which there is insufficient evidence. In particular, G. R. Bode (1958) showed that the allelopathic effect of walnut and black walnut is related to water-soluble leachates from leaves, and root exudates play no role. Similarly, the toxic effects of false flax on flax is linked to the leaching of toxic substances from its leaves, but not with root competition or root exudates, etc. (Grümmer, 1958).

However, it is not only the actual toxicity, but also the mass of release of substances and their ability to retain effective concentrations, as well as their effect on soil microflora, that are of great importance. Therefore, despite the low activity, root exudates in most cases occupy an important place in the overall allelopathic complex of plants.

Excretions from Leaves and Other Aerial Organs

Among the active vital secretions, the most important are guttation, release of gaseous substances and excretion of solid products, which can be leached following precipitation.

The guttational liquid contains from 600 to 2500 mg/l solids, and half of them are organic compounds (Curtis 1944a., Stockling 1956). The most common ions are Ca and NO_3, and also glutamic acid (Greenhill, Chibnall, 1934) which occurs in local necroses of leaves (Curtis 1944b). In guttational fluid, enzymes are found (Stocking, 1956), as are growth substances (Madaus, 1937).

We observed that the intensity of guttation of oats was higher when plants were in a normal day-night cycle. The guttational fluid collected from plants, when lit, contained 14.5 RCU and when the plant was in the dark - 7.2 RCU.

Wheat plants can assimilate penicillin and other antibiotics through the roots and release them with guttational water (Krasil'nikov, 1962). Guttation is used to remove excess salts, especially in halophytes. The secretion of significant amounts of salt, owing to fluid bleeding from branches broken by the wind, and possibly by guttation of fluid, leads to the formation of unique salt marshes under Turanga section poplars (*Populus pruinosa* Schrenk and *P. diversifolia* Schrenk [28]) (Bykov, 1961).

Plants are able to control well the composition of the guttational fluid. For example, in one of our experiments with corn, with P^{32} and Ca^{45}, in the secreted guttational water the total amount of these was approximately half of that in the juice

[28] Now known as *P. euphratica*.

from bleeding. The relationship between these elements is very varied. The ratio of P:Ca in xylem sap was 1:0.28, and in guttational fluid 1:2.3, i.e. in corn, phosphorus is much more actively retained than calcium.

According to our observations, carried out in the steppe reserves and near Kiev (Grodzinskii, Oychnyuk, 1962) guttation is a widespread phenomenon. In nature, 30-80% of Ukrainian steppe species are capable of guttation, depending on the humidity of the habitat.

Guttation was found in representatives of 43 families (except Crassulaceae and Aceraceae). Most prone to guttation are Gramineae, Polygonaceae, Umbelliferae, Labiatae, Cucurbitaceae, Solanaceae; and Leguminosae, Scrophulariaceae and Compositae have less guttational species.

The relative number of species capable of guttation increases from south to north. Of the 35 species, on which observations were carried out in the reserve "Mikhailovskaya Tselina" (Sumskaya region of the USSR), 30 are capable of guttation in nature, and only five were not observed to have this ability. Among the species at Feofania (near Kiev), 117 species were observed as guttating, in 11 guttation was unclear, and in 32 it was not identified. By the way, within the last group, many species were woody, within which guttation is much more difficult to observe than in herbaceous plants. The reserve "Khomutovskaya Steppe" (Donetsk region) guttation was recorded for 122 species, negative results in 85 species.

The overwhelming majority of plants have a low intensity of guttation. Such plants rarely have drops running down, and therefore guttational secretion is unlikely to affect significantly the neighboring plants.

However, there are species in which guttation is very abundant. so that in the morning the whole soil under such plants becomes wet. Of these species should be mentioned field horsetail (*Equisetem arvense* L.) and other species of horsetails, spring sedge (*Carex praecox* Schreb.) purple willow (*Salix purpurea* L.), fig buttercup (*Ficaria verna* Huds.), tomato (*Solanum lycopersicum* L.), species of poppy (*Papaver* sp.), knotweed (*Polygonum aviculare* L.), chickweed (*Stellaria media* (L.) Cyr.) and numerous representatives from the grass family: feather grass (*Stipa capillata* L.), meadow foxtail (*Alopecurus pratensis* L.), common reed (*Phragmites communis* (L.) Trin.), cocksfoot (*Dactylis glomerata* L.) tuber bluegrass (*Poa bulbosa* L.), annual bluegrass (*P. annua* L.), Kentucky bluegrass (*P . pratensis* L.), meadow fescue (*Festuca pratensis* Huds.), awnless bromegrass (*Zerna inermis* (Leyss.) Lindm.), couch grass (*Elytrigia repens* (L.) P.B.), barnyardgrass (*Echinochloa crus-galli* (L.) Roem. *et* Schult.) and others.

Thus, the assertion by A. Frey-Wyssling (1949) that "... upon closer observation it can be established that most of the "dew", moistening our meadows in the mornings, does not fall from the sky, but is secreted by the meadow herbs and grasses", can be considered correct, at least in many cases. This applies not only to meadows; in winter wheat, oats, barley, etc. from tillering until stem elongation, guttation is so intense that it can be compared with a little rain. There is very abundant guttation in rice (Ozaki, 1962).

In some plants, the guttation appears quite peculiar. Thus, in a tulip, droplets appear not only on the tip of the leaf, but also at the middle of the leaf underside

(Figure 8, 3). In spurges, drops on the leaves never appear, and the release of water occurs through the flowers (Figure 8, 4). Generally, as a rule, the lower the leaves are located, the stronger the guttation (e.g., jasmine or rose). In the secretion of cereals, drops through the generative organs are very rare (we saw it only in crested couch grass - *Agropyrum pectiniforme* Roem. *et* Schult.).

In large trees (lime, horse chestnut) guttation is observed at the time of bud break.

The allelopathic importance of guttation can be represented as follows. Drops appearing on the leaves are shaken by the wind onto the ground or lower leaves, and either dry out or soak them. In this way, the transmission of physiologically active substances from one plant to another can occur, and the chemical interaction of plants ensues.

Guttational solutions have a very low concentration. However, their effect is quite possible, because with drying the concentration of solutions increases. In addition, the appearance of the solutions and drying of the leaves is repeated almost every night. According to our calculations daily excretion of organic substances in guttation can be 10-15 kg/ha or more,

As we mentioned already earlier, plants are able to control the composition of the guttational fluid to a wide extent. So guttation is a means of elimination of various unnecessary and even harmful substances. The downward flow of leaf drops can dissolve and rinse new quantities of active substances. For example, it has been shown that in guttation the droplets are able to leach boric acid from living tissues of coleoptiles and cause boron starvation in guttating plants (Oertli, 1962).

It is difficult to discern the allelopathic effect of guttational discharge in nature, because it occurs at the same time against a background of other water-soluble and volatile secretions, which may enhance the effect, and taken separately, it may be too weak. It would seem that guttation most likely plays a role in the relationship of herbaceous plants, since it is very rare in woody plants and its value is more in areas of high moisture, where it occurs often enough and abundantly. It is indeed very important that guttation has maximal intensity in the spring and early summer, when there is strong growth in almost all plants. At this time, the possibility of manifestation of chemical influence is the most feasible because, for example, in the fall, even though there is high accumulation in the falling leaves of cholines, etc., it is ineffectual because in this period the plants have stopped growing and easily withstand any harmful effects.

Great ecological and allelopathic importance is attached to so-called cuticular excreta from the leaves, which appear on the surface as residues and are leached by rainfall, and to substances leached from the living cells of the leaf. Indeed, N. T. de Saussure (1804) believed that the rain washed out of the leaves up to ⅓ of its mineral salts. Newer research has shown that due to some methodological inaccuracies Saussure's data for natural habitats were somewhat exaggerated, but not much, and the loss of substances from the leaves can have serious ecological implications.

Cuticular excretion processes are believed to occur as follows. The water, having come from the soil and containing mineral salts, moves from the ends of vascular bundles in the mesophyll of the leaf, via the cell walls toward the cuticle, from which it evaporates through the submicroscopic pores. Mineral salts are partially absorbed by the protoplasm of cells, and partly deposited in the cuticle and on its surface (Arens, 1934, 1936 b). Hence soluble substances are washed away by precipitation.

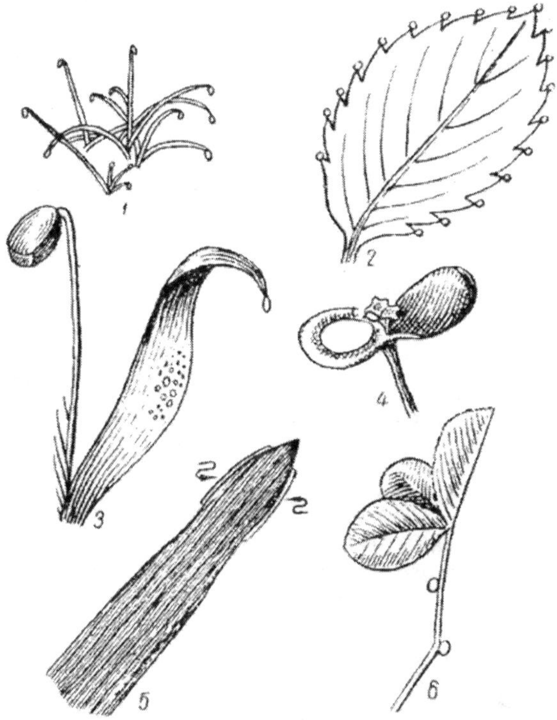

Figure 8. Types of guttation in different plants.
1 – in grass; 2 – in strawberry; 3 – in tulip; 4 – in spurge;
5 – in *Bellevalia sarmatica*[29] 6 – in clover

K. Arens as a result of his research on pondweed (*Potamogeton* L.) came to the conclusion that this process does not occur passively during transpiration, and is due to active mechanisms (Arens, 1936 a). In particular, in the case of an unbalanced supply, "surplus" cations or anions are mostly excreted. The same view was held by Lausberg (1935). D. I. Kolpikov (1960) proved that the widespread solid, liquid and gaseous secretions, whether through stomata and various glands, hairs, etc. were due to actively operating mechanisms.

In contrast, G. Engel (1939) considers the excretion process through the leaves as completely passive diffusion, since, in his view, living protoplasm cannot retain all of its constituent parts. But he does not deny the possibility of the ecological role of these secretions. In his experiments with 24-hour immersion of leaves of 11 species of plants in distilled water, 0.5-5.7% ash content of the leaves was leached.

---

[29] Now known as *B. speciosa*

However, it occurs, it is reasonable to accept both of the mechanisms of formation of solid excretions from the leaves - active and passive.

The excretion of minerals through the leaves can be of very large extent. Using artificial rain, up to 50% of the total ash constituents may be leached without killing the leaf. In the rain or dew water that drips from the leaves, calcium, potassium, magnesium, sodium, phosphorus, and nitrogen compounds are detected. Potassium, sodium, manganese, and other elements that are found in plants in ionic form, are especially easily removed from the leaves (Ratner, 1948). According to Arens and Lausberg, plants during the growing season lose through the leaves more potassium than on average is contained in the crop harvest. So this forms a sort of cycle of potassium between plants and soil. K. Schoch (1955), in studying the cuticular excretion in tobacco and castor bean, found smaller quantities; however, and according to his information, the excretion of potassium from leaves of tobacco still gave removal of this element that was 70% of that of the harvest, and the excretion of calcium - 10%. Leaching of potassium from banana leaves during long tropical downpours leads to deficiency in the plants and reduces crop (Bhan *et al.*, 1959).

Such a strong excretion of salts through the leaves allows certain types of plants to get rid of excesses of these substances.

Also mineral compounds are leached from the leaves, and organic substances that constitute 40-60% of the dry weight are washed from the leaves.

K. Mothes (1938, cited by Knapp, 1954), showed leaching of alkaloids from the leaves of black henbane (*Hyoscyamus niger* L.) and thorn-apple (*Datura stramonium* L.).

In 1950, he established with tobacco and thron-apple, which were artificially sprinkled, that the alkaloid content in them was more than two times lower than that of the plants that were irrigated through the roots. The well-known observation that the content of alkaloids in plants increases in fine weather and decreases in rainy weather may be related to their leaching.

M.N. Moiseeva observed that the number of plant hormones in the leaves of trees after rain decreases, which obviously can be attributed to the leaching of these substances from the leaves. Also, the amount of sugars in the leaves after rain was reduced.

Facts concerning the release of significant quantities of various substances from living leaves by means of precipitation have been reported by a number of researchers (Bode, 1958; Long *et al.*, 1956; Kingsley *et al.*, 1957; Wasicky, 1959; Kirillov, 1964, and others).

Insect damage may give rise to specific secretions on the leaves and aerial parts of the plants. Well-known is "honeydew," which appears after leaf damage by aphids. The outflow of sugary liquid from the phloem sieve tubes, damaged by aphids, may continue for weeks. Huge amounts of sweet liquid are released by the plants themselves and by aphids. Often one can observe the appearance of "honeydew" in June and July, especially during the occurrence of hot weather, on the leaves of maples, lindens, elms, willows, sometimes lilac, jetbead[30]. V.N. Shnitikov (1950) described the accumulation of sugar masses on the branches of the Tian Shan spruce[31], which, in his opinion, was connected with the activity of aphids. Droplets appear as a flow of stringy sweet liquid, and the soil and plants underneath are coated with an adhesive layer, under trees secreting "honeydew". "melliferous dew", otherwise honeydew, contains glucose, melezitose, dextrin, and nitrogenous substances.

The quantity of honeydew is very large. Suffice to say, according to O. Garagsima in Czechoslovakia, that 50-80% of the total honey yield is obtained from honeydew.

Clearly, such a mass release of sugar and other organic compounds cannot but affect the surrounding vegetation.

On July 24, 1963 we have seen in the hornbeam-oak[32] of Feofaniya, under an adult linden tree an area almost devoid of vegetation, in which there were only in a few instances, significant Galeopsis (*Galeopsis speciosa* Mill.), field sow thistle (*Sonchus arvensis* L.) and black bindweed (*Polygonum convolvulus* L.)[33]. These plants were severely damaged, almost without leaves, covered with black sticky mass. Outside the area, that has an outline that corresponds exactly to the canopy of tree, there was luxuriant growth of goutweed (*Aegopodium podagraria* L.). Sticky, black fungal spores from the mass contained a large amount of reducing sugars. Under the three neighboring lindens, there was observed neither "honeydew" nor suppression of the lower plant stratum, which is entirely occupied by goutweed. In 1964, this phenomenon was observed under a nearby tree of linden.

This example indicates a very strong allelopathic effects of honeydew in the event of a mass occurrence. We see that the usual herbaceous component of hornbeam-oak, goutweed, completely disappeared from under the aphid-affected linden trees. and instead weed species increased, and later died under the influence of "honeydew".

Resins or gums consisting primarily of polysaccharides are often formed on the trunks of damaged fruit trees, as well as in some desert astragalus and acacias. In the latter case, their number is so large that it is used as a source of raw material for the production of adhesive (gum arabica, tragacanth, etc.). The so-called "manna", which is formed on the leaves of Mediterranean species of ash, manna tamarisk (*Tamarix mannifera*) and warty euonymus (*Euonymus verrucosa* Scop.) in our latitudes is composed mainly of mannitol and dulcitol [34], but it also contains some physiologically active substances.

---

[30] Jetbead is the occasionally used common name for *Rhodotypos scandens,* and refers to black shiny fruit.

[31] Tian Shan spruce is *Picea schrenkiana*

[32] The term "hornbeam oak" refers to a type of forest vegetation characterised by these two trees, but other species also occur.

[33] Now known as *Fallopia convolvulus.*

[34] Also known as galactitol.

The appearance of the liquid leaf discharge is often caused by the action of insects. Steppe and meadow plants are often attacked by larvae of the cicada-spittlebug.

Our observations made in the reserve "Mihailovskaya Tselina" (Sumskaya Oblast, Ukrainian SSR), where particularly many of these larvae are found, showed that they feed on the sap of more than 40 species of diverse taxonomic position. They are usually located on the younger part of the stems or leaf petioles, and surround themselves with foam-like saliva. The foam bubbles coalesce, and form liquid droplets that flow down the stem to the base of the plant. The amount of this fluid can be very large - on a hot summer day, if walking on the grass in the steppe, where there are many leafhoppers, it is easy to get your feet wet. Liquid formed under the influence spittlebugs is derived mainly from xylem and partially from phloem sap and possesses phytotoxic properties in bioassay.

It is interesting to note that such a "foaming" is observed in the willows, especially in flood plains, above ponds. Foam continuously drops off the branches, and resembles rain. Apparently, this phenomenon was seen in June 1931 on the Dnieper River by Kholodnyi, who mistook it for guttation (Kholodnyi, 1957). He was amazed at the quantity of the water discharge, which reminded him of a heavy downpour.

We are talking here not about the "real" excretion of plants, but as a sort of pathological phenomena. However, it is regularly found in nature and may have some significance to the phytocenosis.

Substances found in the solid form are excreted, obviously in the form of solutions, which then dry out and solidify. In addition to the above-mentioned secretions like "manna", "honeydew", resins and gums confined to certain types of plants, solid water-soluble substances can be found on the leaf surface of almost all species of higher plants. These substances are easily leached from the leaves by rain, and fall onto the plants and the soil below, and, as in many cases, they have a very active influence on vegetation and microflora.

Various extracts, leachates, solutions obtained from the leaves by rinsing, soaking and boiling, phytoncides of pulverised tissues, diffusates of fallen leaves, etc., almost without exception, have very great phytotoxic properties. There are many articles about the allelopathic effect of extracts from leaves or aboveground parts of various plants (Golomedova, 1952, 1954; Gel'man, 1961; Chasovennaya, 1961 *et al.*).

It can be stated more confidently that the extract from the leaves of any plant is physiologically active with respect to many biological objects.

V. G. Drobotko *et al.* (1958) examined the content of antimicrobial substances in 453 species of plants collected from the Botanical Garden of the Academy of Sciences of the Ukrainian SSR, in the vicinity of Kiev, and partly from the North Caucasus. If one considers the antibacterial activity of extracts from these plants with respect to at least one of the test organisms on which they were tested, it appears that antibacterial agents occur in 397 species, representing about 87.8%. The authors suggest that if tests have been more extensive, and plants had been studied in detail, this percentage would have been even higher.

The same can be said about the effect of substances from the leaves of higher plants to other higher plants. It is much harder to find a really inactive plant than one that has a depressive effect on the chosen test object.

In a series of experiments carried out from 8 to 12 August 1962 in the Khomutovskaya Steppe, the activity of leachates from leaves of 42 plant species was studied by bioassays with radish seeds and poppy. The sensitivity of the poppy and radish in these experiments was very similar, the coefficient of coincidence reached an average of 82.4%; in 22 cases the deviation had a sign "+", and in 20 cases "-". The data about the distribution of aggregates of species depending on the activity of the leaves with leachates are shown in Table 16.

Table 16. Activity of water-soluble excretions of leaves from 42 species of plants of the Khomutovskaya Steppe

Class of activity	Germination of seeds at K-50%	Radish		Poppy	
		n	%	n	%
I	0-20%	0	0	0	0
II	20.1-40.0%	2	4.8	2	4.8
III	40.1-60.0%	9	21.4	8	19.0
IV	60.1-80.0%	18	42.9	18	42.8
V	80.1-100.0%	9	21.4	13	31.0
VI	100.1%	4	9.5	1	2.4
	Total	42	100.0	42	100.0

It can be seen that there is good agreement of the data obtained with both test species. The greatest number of plants - 42.9% - yielded leachates that were of moderate biological activity, one quarter of species had a fairly high-level leachates.

Among the most toxic in bioassays were leachates from the leaves of many plants (Table 17).

Table 17. Content of cholines in leachates from leaves of steppe plants

Species of plant	Content of cholines (in RCU)
Russian almond (*Amygdalus nana* L.)[35]	181.0
Common wormwood (*Artemisia absinthium* l.)	122.0
Common toadflax (*Linaria vulgaris* Mill.)	86.5
Great plantain (*Plantago major* L.)	77.0
White mullein (*Verbascum lychnitis* L.)	75.1
Sea lavender (*Limonium latifolium* (Sm.) Ktze.)	75.0
Multi-flower bellflower (*Campanula multiflora* W.K.)	69.5
Redstem wormwood (*Artemisia scoparia*)	60.2

[35] Now known as *Prunus tenella*

In a number of experiments there occurred a noticeable toxicity with leachates from the leaves of burdock (*Arctium lappa* L.), field bindweed (*Convolvulus arvensis* L.), couch grass (*Elytrigia repens* (L.) P.B.) and other plants.

These data suggest that physiologically active substances can often be found in the drops of rain or dew, flowing from the leaves of many herbaceous plants of the steppe zone. There is no reason to believe that the plants are exceptions in the plant kingdom, and that with other plants toxic leachates will occur less frequently.

The release of volatile and gaseous substances from entire intact leaves has long been known and is practically used in medicine or in phytopathology (Verderevskii, 1961); a well known phenomenon is the allergic or toxic effects of some plant odours or pollen to people sensitive to these (Ado, 1961). Most commonly, therapeutic or toxic effects are associated with the release of essential oils and other aromatic compounds or toxic volatile substances, such as with juglone from walnut or clematin from clematis.

N. G. Kholodnyi (1957) has written about the biological significance of phytogenic organic substances in the atmosphere, and he developed the concept of "atmovitamins". Particular attention was paid by Kholodnyi to the influence of phytogenic substances on the microflora, and thus through their mediation, on the growth of higher plants.

There are many facts about the physiological and phytocenotic role of volatile excretions, in particular ethylene, which are provided in the monographs of H. Molisch (1937) and G. Grümmer (1957).

In some very carefully and very technically accomplished work performed by G. A. Sanadze (1961), he has shown that the whole leaves of a large number of plant species are able to release into the atmosphere hydrogen, a number of lower hydrocarbons, aldehydes, and volatile compounds that are acidic or basic in nature.

G.V. Porutskii (1962, 1963) confirmed the data of G.A. Sanadze concerning the emission from leaves of unsaturated hydrocarbons. However, he showed that the secretory function is dependent on the physiological state of the plant. In particular, stimulated plants emit more gases, the composition of which includes relatively higher members of the paraffin and olefin series.

As shown by G.A, Sanadze, the release of volatile phytogenic substances by leaves depends on many external and internal conditions. In particular, less material was excreted from plants grown at 40% humidity, than from those grown at 60% humidity; leaves in August released more substances than at the end of September (in Georgia). No relationship was found between the quantity of excretion and the phytoncidal activity of leaves (relative to infusoria). For many plants that were not distinguished by any odour, upon concentration of the volatiles, they exhibited a very definite smell. Individual plants vary greatly in the number of released substances.

Comparative tests of the gas releasing activity of 27 plant species under identical experimental conditions, as conducted by G.A. Sanadze, have produced results that are displayed in Table 18.

Table 18. Volatile emission potential of leaves, of various plants (after G.A. Sanadze, 1960, 1961)

Species of plant	Total content of determined substances in volume percentage in 1 m^2 of leaf area
Apricot (*Armeniaca vulgaris* Lam.)	0.0019
Ailanthus (*Ailanthus glandulosa* Desf.)	trace
Black locust (*Robinia pseudoacacia* L.)	1.416
Bastard indigo (*Amorpha fruticosa* L.)	1.244
Bamboo (*Phyllostachys* sp.)	0.677
Common privet (*Ligustrum vulgare* L.)	0.097
False hemp (*Datisca cannabina* L.)	trace
White willow (*Salix alba* L.)	2.484
Caucasian hackberry (*Celtis caucasica* Willd.)	trace
Golden rain tree (*Koelreuteria paniculata* Laxm.)	0.117
Common dogwood (*Cornus mas* L.)	trace
Box elder (*Acer negundo* L.)	0.079
European olive (*Olea europaea* L.)	0.046
Dense poa (*Poa densa* N. Troitzky)	0.0013
Holly (*Ilex aquifolium* L.)	0.194
Peach (*Persica vulgaris* Nill.)	0.035
Rose (*Rosa* sp.)	0.132
Common rue (*Ruta graceolens* L.)	0.041
Iberian spiraea (*Spiraea hypericifolia* L.)	0.127
Fescue (*Festuca sulcata* E. Hack.)	0.0014
Chinese poplar (*Populus simonii* Carr.)	2,470
Black poplar (*Populus sosnowskyi*[36] A. Grossh.)	3.536
Pistacchio (*Pistacia mutica* F. ct M.)	0.011
Common hops (*Humulus lupulus* L.)	0.0013
(*Cochlearia armoracia* L.)	0.137
Common persimmon (*Diospyros lotus* L.)	0.737
White mulberry (*Morus alba* L.)	trace

Among volatile excretions of tobacco, Russian peashrub, poplar, willow and other plants, methane, ethane, propane, butane, and a number of unidentified heavier substances were found using gas chromatography and mass spectrometry. Under the same conditions, the evolution of molecular hydrogen was observed.

The results obtained show that there is still a similarity in the chemical composition of the various plant volatiles, which points to the common nature in the occurrence of a number of substances in various species. Between plants there are important differences. So, in poplar after 2-3 hours of exposure to light, hydrocarbons comprise about 25% of substances released. As the residence time in

[36] Now known as *P. nigra*

the chamber is extended, the amount of heavier compounds increases. In willow, on the contrary, these heavier compounds are almost absent. In the secretions of Chinese poplar there is probably the presence of ethyl alcohol.

Concerning the mechanism of formation of volatile phytogenic excreta, G.A. Sanadze (1963) suggests that a connection with photosynthesis can be reasonably assumed. Indeed, their secretion takes place only in the light and is closely dependent on its intensity; photosynthetic poisons (hydroxylamine) not only inhibit photosynthesis, but also the emission of volatile substances. The curve of released substances during the growing season is reminiscent of the curve of photosynthetic activity. The release of gas occurs mainly through the stomata, although the possibility of cuticular excretion cannot be eliminated.

The release of volatiles in hot weather increases and as the temperature decreases, it declines. All sorts of mechanical damage to leaves enhance the release of volatile substances.

Of some interest in assessing the role of allelopathic plant volatile secretions are studies from medical health resorts, which have long shown a significant therapeutic role for plant emissions. The effect of volatile excretions is often associated with air ionisation. The literature on the ionisation of the atmosphere is very extensive; previously ionisation was thought owing to the action of ionising radiation and cosmic particles on molecules of oxygen, nitrogen, etc. Now this phenomenon has also been associated with the activity of the plant ("Aeroionisation and Hydroionisation in Medicine", 1962)[37].

Unfortunately, the study of volatile secretions of intact leaves has not yet attained an appropriate scale. On the other hand, there are very wide-ranging studies on the volatile vapours from macerated leaves, which are called phytoncide production. The nature of these emanations in most cases is not yet known; it is believed that they have strong reducing functions. S. Skvortsov (1960) found volatile aldehydes as part of the phytoncides; the presence of hydrocyanic acid was identified in the phytoncides of wild cherry (Drabkin, 1954), and clematin in clematis (Rudakov, 1956). Allicin, $C_3H_5-\underset{\underset{O}{\|}}{S}-S-C_3H_5$ has been isolated from phytoncides of garlic; chemists have prepared a number of its synthetic analogues, called pseudoallicin, which have similar physiological properties. Phytoncides from the plants of wheat proved to be 6-methoxybenzoxazolinone, and in rye-benzoxazolinone.

B.P. Tokin (1960) believes that the phytoncides are substances that enhance the immune system of plants, and therefore they are common to all plants. Phytoncides according to B.P. Tokin, should be regarded as products that are formed in certain organisms and which have a harmful effect on other organisms. Phytoncides are usually regarded as involving volatile vapours that arise in damaged plant tissues.

---

[37] See Vasil'ev 1962.

Most of the results on the detection of phytoncides have been obtained using as a simple bioassay, with paramecium and infusoria, or microbiological objects.

Studies to elucidate the action of phytoncides on higher plants were conducted by A.A. Chasovennaya (1954, 1955). In her experiments, using Petri dishes on a special table, she placed a mash from macerated leaves in a dish, and in the same dish on wet filter paper, the seeds of studied species. The author found a significant effect of volatile compounds from the pulverised mass on seed germination and initial growth of the plants. A.A. Chasovennaya very carefully examined the physiological effects of the phytoncides of hogweed (*Heracleum mantegazzianum*) and a number of steppe plants. Similar studies with pasture grasses were conducted by A.A. Gel'man (1961).

We conducted a series of studies on the phytoncidal action of steppe plants, using a bioassay to quantify their activity with the seeds of radish, lettuce and poppy, and to allow comparison with the data given in the literature, which used paramecium in bioassay.

The phytoncidal activity of most plants varies in small degree during the day, and the fluctuations are not characterised by any regular pattern: sometimes the maximal activity was observed in the morning, and sometimes at the end of the day.

Leaves from different layers of the plant have differing phytoncidal activity. More than half of the plants studied have phytoncidal activity that increases regularly from the bottom to the upper leaves. Generally the uppermost leaves have the greatest activity compared with the other organs of the plant in 87.5% of cases. This pattern seems to be general in nature; the younger organs of plants always have phytoncidal activity that is greater than that of similar, but older structures. In the same way, young plants are more phytoncidal than older ones, which have grown in similar conditions and collected at the same time. Plants at the stage of budding and flowering are more phytoncidal than vegetative plants.

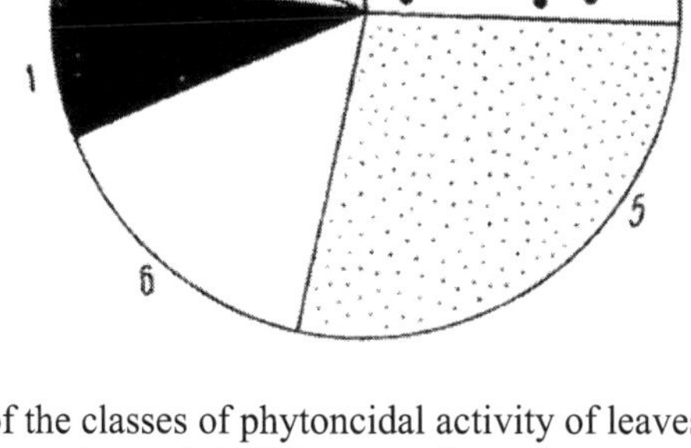

Figure 9. Piechart of the classes of phytoncidal activity of leaves from 238 species of plants of the Khomutovskaya Steppe:
1 – toxic; 2 – highly active; 3 – average activity; 4 – slightly active;
5 – inactive; 6 - stimulatory

Data concerning the analysis of leaves of 238 plant samples representing more than 150 species of herbaceous plants from the Khomutovskaya Steppe were analytically treated, and are shown in Figure 9.

In analysing the data presented in Figure 9, one can see a fairly large number of facts concerning significant inhibition and significant stimulation. Thus, there were 19 samples, or 8% of the total, giving the highest activity of 100 relative units; 9 samples, or 4.2% of the total, with an activity of 50 to 100 units; and 15.8% with activity 30 to 50 units - and in all 28% of the total number samples showed credible inhibitory activity.

Weak inhibiting activity (from 10 to 30 relative units) was also found in 29% of the samples, an additional 28% were inactive in dosages used, and finally, 16% of the samples stimulated the germination of radish.

The most active phytoncides were found in dwarf Russian almond (*Amygdalus nana* L.), false fragrant virgin's bower (*Clematis pseudoflammula* Schmalh.), erect clematis (*C recta* L.), wormwood broom (*Artemisia pontica* L.), licorice and wild licorice (*Glycyrrchiza glabra* L., *G. echinata* L.) catnip (*Nepeta cataria* L.), sage (*Salvia officinalis* L.) and tansy (*Tanacetum vulgare* L.).

Based on these data, one cannot establish any definite connection between the phytotoxicity and the systematic position of plants. It seems unlikely that it exists in reality. The activity of volatile substances from six species of wormwood varied widely., -9 to +375 conventional units.

Similarly, representatives of grasses can be found throughout the range of values, along with representatives of other families. There is also no connection between phytoncidal activity and human notions about the smell.

It is of interest that the stimulation of seed germination of radish and other plants, under the influence of phytoncides may reach unusual values: 145% compared with the control (170% in one case), which is almost never observed with the effect of water-soluble stimulants.

Phytoncidal activity towards higher plants, protists and microorganisms is of varied nature. Only in rare cases is the same plant toxic to all of these organisms (e.g. almond and wild cherry, thanks to hydrocyanic acid). In most cases, the phytoncides of the plant cause the death of protozoa or bacteria, but has little activity against other higher plants. Thus, the phytoncides of onion and garlic, are destructive for paramecium, but their effects on seeds are weak or even cause stimulation of germination. Therefore, data obtained from microorganisms or protists, cannot be transferred to relationships concerning higher plants.

Phytoncides are active not only with respect to germinating seeds, but they also affect the growth of roots, discs of leaves, photosynthesis, guttation and other manifestations of plant life.

As we have noted above, the phytoncides of roots are usually ten times weaker than the phytoncides of leaves. The volatile emanations of crushed stems occupy an intermediate position, i.e., if the roots mostly stimulate the germination of seeds, then the leaves are suppressive, and the stems do not affect this process. The phytoncidal activity of macerated flowers usually equals the activity of the upper leaves.

Seeds of different plants respond in varying degrees to the phytoncides of macerated tissues. Thus, according to V.V Arendarchuk, lettuce appeared to be more sensitive than radish or poppy, which generally responded similarly to phytoncides. This can be confirmed by the frequency distribution plots of the activity of phytoncides determined using a bioassay with seeds of the three species (Figure 10).

In the literature, there is little information on the impact of odours of flowers on plant subjects. A number of observations are given in the book by S.I. Chernobrivenko (1956), mainly concerning the behaviour of plants in clumps. Analogous results are also given in the book by Grümmer (1957), who, however, does not give them much value, since the issue is more about the action of water-soluble substances than volatiles.

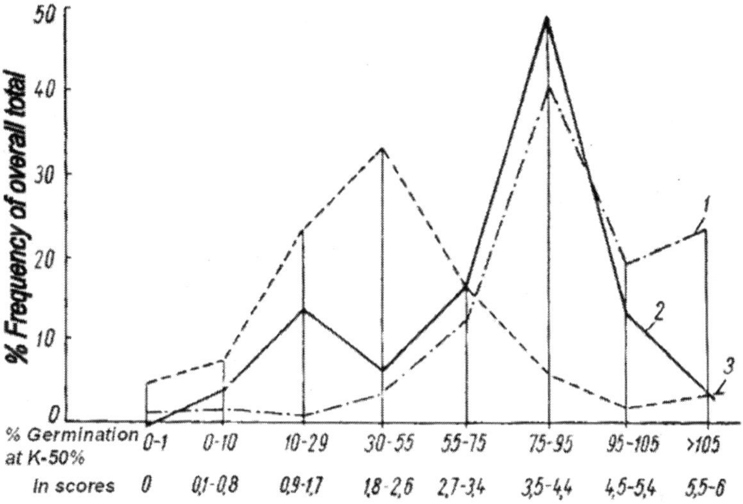

Figure 10. Histograms of frequency distribution of the active phytoncides determined using bioassays on the seeds of radish (1), poppy (2) and lettuce (3).

In the literature there are indications concerning the release of vitamins from flowers (Kakhidze, Medvedeva, 1956). According to A.N. Sverchkov (1964), flowers are powerful centers of forming air ions, and he noted many differences between individual plants. The author found a very interesting phenomenon. It was found that with increasing concentration of volatiles compounds, the number of negative light ions initially increases, and then declines as the concentration of volatiles further increases. At the same time the number of heavy negative ions increases, and all the electrical parameters are unfavorable from a medical point of view. This phenomenon, which is monitored physically using the device, an air ion counter, is explained by the known physiological phenomenon: the stimulation of life activity occurs with small doses, and depression by large doses. So far as we know, these particular physiological effects have found no physical or chemical explanation.

We conducted several experiments on the effects of the odours of flowers of 17 plant species by bioassay with radish. The experiments lasted from 6 PM to 9-10 AM of the next day. The results of these experiments showed that the fragrance of flowers significantly affects the germination of radish, with inhibition or stimulation occurring according to concentration. Activity of the 17 species investigated varies. The effect of volatile secretions from the flowers on seed germination does not correlate with human notions about the smell, if we compare the different types. Thus, quite a strong scent such as phlox causes stimulation of germination, but the scent of thorn-apple and sweetclover, also strong, caused almost no effect on seedling emergence.

Within a single species, the strength of the physiological effect is positively correlated with the subjective perception of its smell.

Despite the fact that in these experiments, there has been found a noticeable effect of phytoncides from flowers on physiological processes in higher plants, this phenomenon in nature is unlikely to play an independent allelopathic role. Secretions from the dying parts of above-ground organs and leaves, apparently, are most important for allelopathic relations in natural phytocenoses where large amounts of organic litter accumulate. In practice, dying remnants contain all the substances that are generally found and revealed in plants and which are too numerous to list. Therefore, it is understood that primarily there are toxic effects of aqueous and other extracts from the leaves on other plants.

There are other possible secretions of plants that are still poorly understood. Among them, we should remember pollen and nectar secreted in flowers. Extrafloral and intrafloral nectaries attract insects by secreting a sweet liquid containing usually glucose, fructose and some minerals. The productivity of nectaries is assessed by beekeepers, and reaches, such as in buckwheat and phacelia, values of 300-600 kg/ha, and in false acacia[38], 1700 kg/ha.

After flowering, it is possible that the nectaries resorb substances that were not used by insects (Lüttge, 1962). These glands are connected directly to the phloem. Yet, probably most of the nectar is washed away with rain and falls onto the soil with the perianth, with which it is possible that there are chemical effects on other plants.

Pollen that is produced in large quantities by wind-pollinated plants, contains large amounts of vitamins, auxins and other physiologically active substances. There are numerous cases of "hay fever", bronchial asthma, and conjunctivitis caused by pollen. Pollen has a major chemical effect on the germination of pollen of other species, on seeds, on parts of the flower (Golubinskii 1946, 1951).

Summing up this brief review of the excretory functions of plants, we can conclude the following.

The plant, throughout its development from seed to natural death, is continuously releasing into the environment a variety of water-soluble and volatile organic products. The total amount secreted is commensurate with the size of the yield of the plants.

---

[38] Also known as black locust, this is *Robinia pseudoacacia*

The intensity of the secretion of substances usually increases under the influence of unfavorable conditions and occurs in especially large amounts in withering plants.

Based on a systematic study of various types of secretions from a large number of plant species, it can be concluded that physiologically active substances, as detected using the method of biological tests, are very widespread in the plant kingdom. Almost no plants - we find it difficult to name any – have secretions which have not been found to be toxic to some extent.

It can be assumed that during the lifespan of active secretion in plants that, on average, ⅓ of species have very toxic secretions, ⅓ less toxic, and the secretions of ⅓ of species do not significantly affect the growth of plants, and even encourage it.

The distribution of species based on the activity of cholines in passive secretions arising from a particular tissue injury or by the leaching of substances from precipitation is respectively 50, 35 and 15%. As for the post-mortal secretions that we propose to call miasmins (volatile) and saprolins (water-soluble), the toxicity of these substances is spread even more widely. According to B. Rademacher, the most important are "... substances, released in a modified or unmodified form from dead plants and their parts and playing a much greater role in the mutual influence of plants than substances secrete by living plant matter" (Rademacher, 1959). In the cenosis, the effects of all types of excretions are cumulative.

As a rule, the greatest toxicity occurs in leaves and their excretions; the smallest in the roots and root excretions; stems occupy an intermediate position. There is a definite correlation between the youthfulness of tissue and activity of the excretions, between protein content and toxicity, etc.

Reported studies on allelopathy in cases of toxicity of some plants (wormwood - common and Austrian, walnut and black walnut, encelia, guayule, couch grass, etc.) are not isolated. They are the rule rather than the exception in the vegetable world, so allelopathy is not accidental, but a natural, widespread phenomenon of importance for the life of plants and their communities.

Physiologically active substances, excreted from plants, or formed from inactive substances as a result of microbiological and other transformations, together create around the plant definite biochemical areas, of which the concentration and activity depend on many factors.

Our knowledge about the nature of plant excretion is still very incomplete and in need of further understanding.

Chapter 3

ACTIVE SUBSTANCES AND THEIR ACCUMULATION IN THE ENVIRONMENT. THE PROBLEM OF ENVIRONMENT

Essential for allelopathic interactions of plants are the physico-chemical properties of the active substances, their capacity to withstand the destructive action of external factors in their capacity to move and accumulate in the environment. Within the overall cycling of physiologically active substances, which is in itself allelopathy, in this section we include the means by which a substance moves from one plant to another.

Information on the chemical nature of the active substances is very scarce. There are even fewer data about the accumulation in the environment of certain physiologically active compounds. Therefore, in the individual chapters of this book we have to talk about the overall physiological effect of cholines of unknown chemical nature.

Substances involved in the chemical interaction of plants

Despite the large number of facts revealed about the chemical interaction of plants, there are only a few cases concerning known substances, that have favourable, or more often, toxic effects. The identification of natural organic compounds is often beyond the power of a biologist, and to do this work it necessary to involve chemists.

Moreover, in nature compounds do not normally occur individually, and there are indeterminate mixtures of different substances, where the activity of each compound has a very complex dependence on many factors. That is why attempts to reproduce an allelopathic effect with the use of synthetic drugs have often ended in failure. Some researchers believe that the problems of chemical interaction can be studied only with the knowledge of the chemical nature of the active ingredients. Without denying the importance of knowledge of the chemical nature of the active compounds, we note only that such an approach will take a very long time to make research in the field of allelopathy more rigorous.

Of course, one must pay more attention to the identification of natural physiologically active compounds, but at the same time it is necessary to conduct research, and with unknown mixtures, using the method of biological assay for evaluation.

In carrying out the analysis of various plant excretions by means of biological assay, each time we determined the presence of different groups combined in the excretions, and then we established the existence of a correlation between the strength of inhibition of growth and the presence of certain substances. In addition, we tested the effects of different chemicals with the same bioassays, and in such a way validated the findings reached.

Based on these data, as well as information gleaned from the literature, we believe it is possible, that all substances involved in chemical interactions of plants, can be divided into three large groups. The first of these substances are those of so-called secondary origin - organic acids, essential oils, alkaloids, glycosides, flavonoids, tannins and other polyphenols, vitamins, antibiotics and phytoncides, and others. These substances are formed as products of normal metabolism and have are comparatively strictly confined to species. The production and release of such substances may have occurred in the course of evolution, as a protective device against herbivores, insect infestation, fungi, bacteria (Fraenkel, 1959), as well as a means of affecting other plants by allelopathy.

In adopting the concept of multifaceted substances, whereby a substance in the organism can perform several functions, one can assume that each secondary substance is intended biologically to protect the organism from different influences simultaneously.

These substances are accumulated and produced either in each plant cell, or in special compartments. Their secretion and impact occur during the life of the plant and after withering away. We emphasise once again that there is great species specificity in the formation of these substances; sometimes the composition of matter of secondary origin not only differentiates individuals of different species, but even different physiological races within a species. The build-up of secondary substances seems to be a good taxonomic marker.

The second group of allelopathic active substances is formed by hydrolytic and autolytic decay of plant proteins in the normal course, but more often during disturbed metabolism and death of tissue. With this decay are formed numerous fragments of protein molecules, often "irregular" structures: peptides, amino acids, acid amides, and amino and imine derivatives, indole derivatives etc., and finally ammonia. These substances are consistently found in plant secretions and, as shown by bioassays, some of them are extremely toxic to plants. By analogy with the so-called "cadaveric poison", they are products of post-mortal decay of proteins that could also be very harmful to the plants themselves. Clearly these materials can be formed in any plant tissue containing proteins. As a rule, the organs of plants that have more abundant proteins, for example, those that are younger, generate and release more active inhibitors.

Such substances have strict specificity with regard to the species donor; they, apparently, are generated and their activity is soon determined by the conditions in which the protein decomposition occurs.

The third group of substances is of very great phytocenotic importance due to their high stability under natural conditions, and the variety of products of the more or less thorough mineralisation and humification of plant material, the so-called humic acids and the whole range of their precursors. These – complex compounds, often are polymeric structures, mainly aromatic, often containing nitrogen heterocycles, characterised by the presence of phenolic and quinone groups. They often have the property of forming complexes with metals (chelates). These substances are formed during the bacterial or fungal decomposition of carbohydrates (cellulose), lignin and other compounds. Their high physiological activity is well known. The problem of humus is, of course, a separate scientific problem, but in some ways, it is so closely intertwined with allelopathy that it would be unforgivable to take from the balance sheet such a large group of substance just because they are not directly among the products of higher plants, or because these substances are traditionally studied in isolation from the problem of the interactions of plants, except perhaps, in the book by Kloke (1963) and some other works.

Considered typical and important for our problem especially are representatives of these groups of active substances.

Organic acids. In the plant world, this is a very common class of compounds characterised by the presence of one or more carboxyl groups. They are important intermediates in metabolism; in many plants they accumulate in significant quantities in the cell sap or in various tissues in the form of salts of calcium, magnesium, iron and others, along with alcohols or ethers. They are usually soluble in water, and the lower homologues are volatile, which ensures rapid migration in the environment of the phytocenosis. Organic acids are starting materials for photosynthesis and the dark reactions, moreover, they are used for regulating the electrostatic balance of the inside of the plant, and between plants and the environment.

Organic acids are formed with great ease and differ among plants. They are found in root secretions, in guttational liquid and solid leaf secretions, and in the gaseous emissions of various organisms.

Especially a lot of organic acids are released upon death and decomposition of plant material. Many organic acids are waste products of fungi and bacteria, fermented from carbohydrates.

Organic acids are often present in secretions in the free form, which causes a shift in the concentration of hydrogen ions in the aqueous solution to the acidic side. This cannot but affect the physiological activity of other substances, dissociation, and activity that is dependent on the pH.

Activity of polar substances (with acidic or basic character) depends on the degree of dissociation of their molecules, which is determined at a given concentration, by the reaction medium (Dykyj-Sajfertova, Dykyj 1956). Some substances are active as undissociated molecules, others, on the contrary, as ions. Also, with acidification of the solution, the permeability of the protoplast increases, which contributes to an easier penetration of substances into cells and organs of plants. Finally, under the action of organic acids, hydrolysis of complex compounds such as glycosides, may occur and may split off the active part of the molecules, or vice versa, the activity may disappear.

In various natural solutions that are involved in chemical interaction of plants, the pH ranges from 3.0-4.0 to 8.0-9.0, often persisting at 5.5-6.5. Naturally, with a different pH, the action of the same compounds will vary.

At extreme limits of pH, the physiologically active effects are stronger.

The physiological effect of organic acids is due to the fact that they are of utmost importance in plant metabolism, or their immediate analogues, have a high chemical reactivity and lend themselves to a variety of metabolic processes.

Some of the dicarboxylic acids (succinic acid, fumaric acid, and others), according to V.P. Filatov (1951) and A.V. Blagoveshchenskii (1962), are biogenic stimulators, i.e. substances of a nonspecific nature that are generated by a variety of plant or animal tissue, separated from the body and produced under adverse conditions, such as cold. These substances have a stimulating effect on other living organisms in contributing to the regeneration of normal functions and improving the livelihoods of organisms.

To some extent, biogenic stimulators are similar to so-called necrohormones, wound hormones that are generated during the decay of damaged tissue and promote the division of resting cells and healing units. One - traumatic acid - is 1-decene-1,10-dicarboxylic acid.

Numerous trials of biogenic stimulants on plants in experimental and production environments have shown good results. Thus, pre-treatment of seeds with 0.002-0.005% strength succinic and 0.02% strength nicotinic acid gives a significant increase in grain yield, and cotton and vegetable production (Blagoveshchenskii, 1962; Polyanskaya *et al*., 1963).

Acid cyclical structures, such as benzoic, shikimic, chlorogenic, caffeic, cinnamic, salicylic, coumaric, melilotic, protocatechuic, and gallic acids, and others have in their molecule one or more rings and are not found in all plants, i.e., those having definite species specificity, as mentioned, tend to have high physiological activity.

Leaves and roots of cotton in particular are activated by the products of anaerobic decomposition of the roots of alfalfa, which are dominated by organic acids (Semergei, 1951).

In the secretions of plants sometimes is found an inorganic acid - hydrocyanic acid HCN. It is a component of amygdalin glucosides, linamarin, and is liberated when split by emulsin or by other enzymes. Hydrocyanic acid is found in fruits and leaves of bitter almond (*Amygdalus nana* L.), common peach (*Persica vulgaris* Mill.), bird cherry (*Padus racemosa* (Lam.) Gilib.) and other Rosaceae; under certain conditions, it is formed in the stalks of sorghum, in flax, and probably many more plants.

The acid is definitely toxic to plants, animals and microorganisms. Many say HCN has a toxic effect on seed germination.

In our experiments, it very strongly inhibited photosynthesis, causing the death of leaves and roots. However, it was toxic to varying degree in all plants: both those that accumulate its derivatives, and those that do not contain it. It is the main component of the phytoncides of wild cherry. However, the allelopathic value of hydrocyanic acid in nature is limited due to the fact that it is usually found in plant tissues in bound form, and after liberation, it quickly disappears.

In general, one should note the crucial role of allelopathic organic acids. Most of them are relatively rapidly degraded in the environment, but due to the constant replenishment level of organic acids can reach phytotoxic values.

Lactone acids. Oxyacids form an intramolecular linkage between a hydroxy group and a carboxyl group with a loss of the water molecules, so that there are compounds known as lactones. The most famous inhibitors of this group of compounds are coumarin, its derivatives and their glucosides as well as a number of others close to them, in having the structure of unsaturated lactones, such as parasorbic acid, aesculetin, ayanin, bergenin, anemonin, cichoriin, daphnetin, nodakenin, nodakenetin, pimpinellin, sesalin, skimmin, umbelliferone, etc.

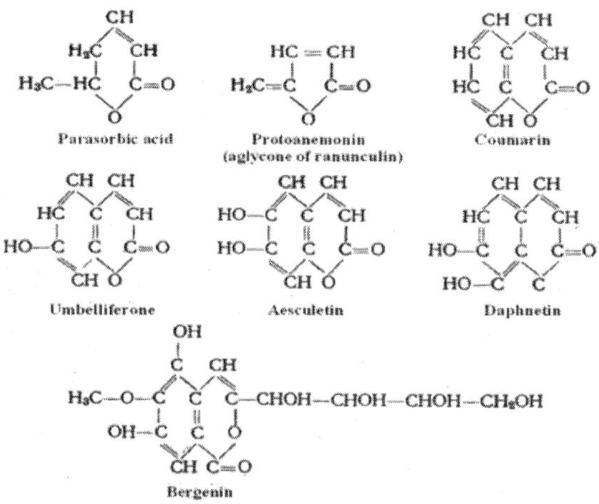

These substances not only delay the germination of seeds and pollen and root growth, but also inhibit the development of fungal, microbial and animal cells. Blaim (1955) found that coumarin, without preventing swelling of wheat seeds, delayed the flow of water into the embryo. It is believed that coumarin is an antagonist of indoleacetic acid – it is an antiauxin in plants.

Tronchet (1962) believes that all lactones, including coumarin, disrupt the polarisation of cells in growing tissues, and because of this disruption, they then stop growing.

Coumarin and other lactones may also act as antagonists of vitamin K and play an important role in photosynthesis.

Unsaturated lactones are widely distributed in nature. Coumarin is found in sweet clover (*Melilotus* Adans.) sweet grass (*Hierochloa odorata*[1] (L.) Wahlbg.), sweet woodruff (*Asperula odorata* L.), in a variety of other plants, particularly grasses, orchids and dicots, in ferns (species of the genus *Adianthum*). It is found in the sap of maize.

The vapours of coumarin in our experiments resulted in average inhibition of germination of radish.

---

[1] Now known as *Anthoxanthum odoratum*.

Aesculetin is found in the bark, fallen leaves and pericarp of horse chestnut (*Aesculus hippocastanum* L.). It also turned out to be very active in its pure form and in extracts. Daphnetin and umbelliferone are found in the bark of wolf bast (*Daphne mezereum* L.) - a poisonous plant. Umbelliferone is found in large amounts in the resin of Umbelliferae. Parasorbic acid is part of the oil of mountain ash (*Sorbus* L.), and a lot of it is in the ripe fruits of mountain ash.

Many moulds form and secrete unsaturated lactones. These include patulin produced by *Penicillium notatum* and other types of *Penicillium* and *Aspergillus*, and also penicillic acid is produced by *Penicillium notatum.*

Obviously, lactones being very strong inhibitors in some cases are responsible for the allelopathic effect of plants, such as clover or horse chestnut. As one knows, only a few adapted species can exist under horse chestnut trees. Lactones are relatively standard and can accumulate in the environment.

Alkaloids – These are nitrogen-containing plant bases, arising in part in protein metabolism, and partly in the other stages of metabolism, that have a strong influence on the nervous system of humans and animals. In their pure form, they are poorly soluble in water; in cell sap they are present as salts of organic acids. They occur mainly in certain families of dicotyledons (Solanaceae, Ranunculaceae, Papaveraceae, Apocynaceae, Leguminosae). They are unlikely to occur in plants rich in essential oils (Labiatae, Compositae).

According to some reports alkaloids are produced in the roots and move to the aerial parts (nicotine in tobacco). However, this does not apply to all alkaloids, such as in lupine alkaloids that are synthesised in all organs. Alkaloids are chemically very diverse. Frequently in addition to the main plant alkaloid found, there are several different compounds, but with similar structure. Over 800 alkaloids have been described.

Alkaloids are often credited with allelopathic effect. However, as noted by G. Grümmer (1955, 1957a), alkaloids do not have a strong effect on plants. In particular, many of them have very little effect on germination.

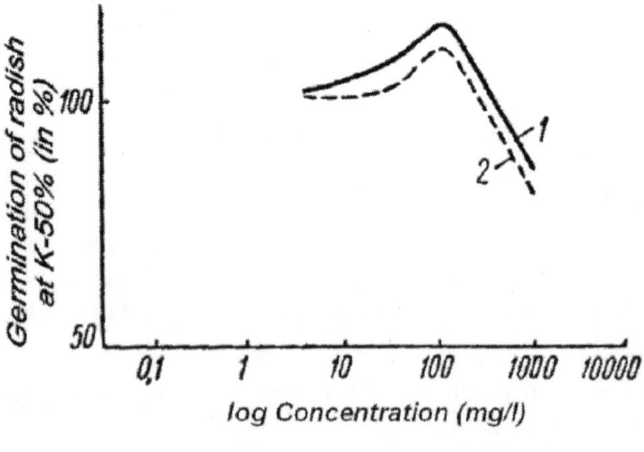

Figure 11. Effect of scopolamine (1) and hyoscyamine (2) on germination of radish depending on concentrations.

Figure 11 shows the curves of action of solutions l-hyoscyamine and scopolamine hydrobromide, which are almost identical. These substances inhibit germination weakly in high concentration (about 1000 mg/l) and noticeably are stimulatory when they are diluted.

Other members of the alkaloids are found in the buttercup family - clematin, anemonin, protoanemonin etc., and possess a very strong antiphytotoxic reaction. In blocking them, the effect is probably due to the blocking action of sulphydryl groups, as when adding cysteine or glutathione, making it possible to weaken the toxicity (Putrimas 1963).

Of particular importance are the alkaloids that affect cell division and, in particular, cause polyploidy, for example, colchicine from the autumn crocus (*Colchicum autumnale* L.) and podophyllin from *Podophyllum peltatum* L. Given that colchicine can fairly easily be leached out via precipitation, one might assume that autumn crocus has an allelopathic effect on its nearest neighbours, but we have not encountered any signs of this in the literature.

In general, it can be said about alkaloids that they probably play only a minor role in direct allelopathy. Only for plants such as Adonis (*Adonis* L.), can we assume a direct impact on neighboring plant due to alkaloids. Other members of the Ranunculaceae release alkaloids only when mechanically damaged, and under natural conditions, very many plants coexist perfectly with them. The same applies to the Solanaceae, Papaveraceae, and Leguminosae. However, the indirect effects of plants, via the microflora and fauna, by means of alkaloids cannot be overestimated.

Glycosides are produced from a sugar which is connected by an oxygen link to the non-sugar (aglucone or genin). According to the type of sugar involved, these are called glucosides, fructosides, rhamnosides, galactosides, etc. Aglucones can be involved in various compounds: aromatic rings (salicin, fraxin, aesculin, arbutin, amygdalin), sulphur mustard oils (thioglucosides), steranes (digitalin, erizimin, saponins), polyphenols (tannins and other tannic substances) or pigments (anthocyanins, flavonols). Among glucosides are vital substances such as riboside adenosinephosphoric acid, vitamin B_{12}, nucleic acids, etc. Glucosides are often redundant substances or the products of neutralisation of toxic metabolites. They certainly play an important role in the natural plant immunity (Rubin, Artsikhovskaya, 1960) and in allelopathy.

Due to the huge diversity of glucosides, there can be found different physiological activities among these substances.

One might think that the activity of many glucosides often manifests itself after hydrolysis. The roots of peach secrete amygdalin (Patrick 1955), which itself is harmless to plants, but after cleaving of glucose, form hydrocyanic acid and benzaldehyde, which mainly determine the phytotoxicity of this glucoside.

The thioglucoside glucobrassicin, which is found in the Crudiferae in representatives of the genera *Brassica*, *Raphanus*, *Nasturtium*, has a significant phytotoxic activity. Moreover, as noted by M. Kutaček (1961, 1963), glucobrassicin acts not on all plants; although it appears to be a potent inhibitor: it is harmless to plants, which are unable to hydrolyse it, or vice versa, it is hydrolysed to harmless products (cabbage, radish). Clover and wheat, which split glucobrassicin in half - to indoleacetonitrile - suffer from this glucobrassicin according to M. Kutaček, in that it gets into the soil with the post-harvest residues of cabbage, radish, rape, etc., and can affect the subsequent crop.

Glucosides, which are contained in the fallen leaves of willows, poplars, ash, horse chestnut and other trees are rapidly destroyed, leaving highly active aglucones. The latter is what determines the significant phytotoxicity of the litter of these plants.

Saponins – This is a group of glucosides, distributed mainly within the pink family, as well as in bog arum (*Calla palustris* L.). They greatly alter the surface tension to form a foam like soap and dramatically affect the permeability of the cells. Saponins can dissolve red blood cells.

Just as with alkaloids, glucosides probably play a more important role in indirect allelopathy – they alter the microbial population of the phytocenosis. Owing to their high solubility in water, they migrate within the cenosis; because of this, they are easily hydrolysed in the environment of the phytocenosis, which does not form an accumulation of glucosides, as well as their more stable aglucones.

Tannins and polyphenols. These represent a fairly heterogeneous group of substances, which are widespread use in the plant world. They have a bitter taste, and they precipitate proteins and alkaloids. It is known they undergo a colour reaction with iron salts in a neutral medium and oxidation in an alkaline medium to a brown colour, and are slightly soluble, which indicate substances that are of polyphenolic nature. Tannins chemically are esters that are a complex of sugars (glucosides) and in most cases tannic acid.

A review of the distribution and the role of some polyphenols in plants is given by E. Unger (1961).

The orthodiphenol, catechol, is a rare component in plants. It has been found in the developing buds of willow and poplar, in plane tree, horse chestnut and physalis (*Physalis alkekengi* L.), and in the root secretions of lovegrass *Eragrostis curvula* Nees. Methyl esters and derivatives of allyl catechol are fairly widespread components of essential oils.

The paradiphenol, quinone, is very common in the plant world. It is found in the free form and in the glucoside arbutin and meloarbutin in the leaves, but less in the fruits, flowers and bark of numerous representatives of the families Saxifragaceae, Ericaceae, Rosaceae, Pyroleae, and Rubiaceae. Ethyl quinones are a component of oils.

The symmetrical triphenol, phloroglucinol, is very rare in the free form, but is regularly present in the decay products of more complex phenolic compounds such as tannins, flavonoids, and anthocyanins. Phloroglucinol is also a product of the decay of phloretin – the aglucone of two related glucosides: phlorizin – a component of the bark and roots, and to less extent in the leaves, buds and fruit trees of the genera of plum (*Prunus*), pear (*Pyrus*) and apple (*Malus*), and glyciphyllin – a component of the leaves, petioles, flowers and seeds of plants of the genus sarsaparilla (*Smilax*). Phlorizin and phloretin are released during the decomposition in the soil of roots of apple tress, and are the cause of growth inhibition and death of seedlings planted in places of uprooted old trees.

The glucoside of the dimethyl ether of phloroglucinol, taxicatine, is found in the leaves and branches of yew (*Taxus* L.).

Other isomers of triphenol, pyrogallol and hydroxyquinone, have not yet been found in plants.

The monocarboxylic phenolic acids, gallic (3,4,5-trihydroxybenzoic) acid and protocatechuic (3,4-dihydroxybenzoic) acid, are widely distributed in the plant world, mainly in the form of esters or glucosides.

Gallic acid is a normal component of hydrolysable tannins and sometimes condensed tannins. It is the aglycone of glucogallin from the roots of rhubarb (*Rheum* L.). Free gallic acid is found in the leaves of sumac (*Rhus*) and tea (*Thea*), in the bark and roots of the pomegranate (*Punica granatum* L.), and in particular, where there is a high content of tannins.

Protocatechuic acid, in free or bound form, is found in the leaves, fruits, wood, barks, flowers of many angiosperms. Free protocatechuic acid is found in the scales of onion (*Allium cepa* L.), star anise fruits (*Illicium* L.) and leaves of grapevines (*Vitis* L.). It is part of the tannins from leaves of apple trees, bark of oak, bark and flowers of acacia (*Acacia* L.). The 3,4-dimethyl ether of protocatechuic acid, the so-called veratric acid, is widely distributed.

Para-hydroxybenzoic and vanillic acids are found in the excretions of couch grass (Grümmer, 1961).

Hydroxy derivatives of cinnamic acid are widespread and have high physiological activity. Caffeic (3,4-dihydroxycinnamic) acid is isolated from clematis (*Clematis vitalba* L.), chamomile (*Anthemis nobilis*), spotted hemlock (*Conium maculatum* L.), tobacco (*Nicotiana* L.), opium poppy (*Papaver somniferum* L.), and cinchona bark (*China cuprea*)[2]. It occurs in the Orobanchaceae – as the glucoside in broomrape (*Orobanche*).

Derivatives of caffeic acid are widespread - chlorogenic acid is found in the seeds of coffee foremost, and then in the leaves of a number of higher plants. It is found in the free form and in its 3-methyl ether - ferulic acid (3-methoxy-4-hydroxycinnamic acid), which is found in the juice of tomato fruits.

Sinapic acid or 3,5-dimethoxy-4-hydroxycinnamic acid is an ether of gallic acid[3] that often occurs in cruciferous plants.

Under the influence of polyphenol oxidases and peroxidases of polyphenols, phenolcarboxylic acids and certain flavone glucosides form quinones that have phytotoxicity and are often volatile (Skvortsov, 1964).

It has long been known that the allelopathic effect of walnut (*Juglans regia* L.) and black walnut (*J. nigra* L.) is attrubuted to quinones - juglone, which is formed by the oxidation of hydrojuglone contained in their leaves:

---

[2] "China cuprea" is an old pharmacological name for a type of medicinal bark that yielded some quinine. It was not from China, but originated from the Colombian trees, *Remijia pedunculata* and *R. purdieana*.

[3] Sinapic acid is a phenylpropanoid compound (C_9), and this it is highly improbable that it is derived from gallic acid, which is a simple benzoic acid (C_7). It is derived from cinnamic acid, and intermediates such as *p*-coumaric, caffeic and ferulic acids.

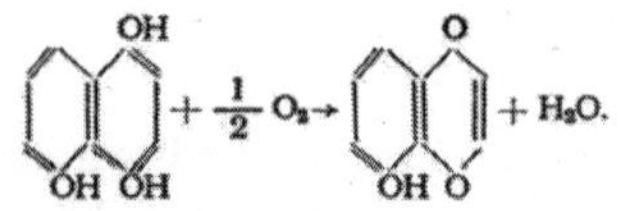

Juglone is washed out of the leaves of the walnut by precipitation and poisons herbaceous vegetation located within the projection of the crown.

Tannins and phenols are easily removed with water, especially if boiled. They enter the environment of cenoses in huge quantities via the litter of woody plants. The presence of tannins in the litter is the reason that the microflora of forest cenoses is represented mainly by fungi, as the bacteria cannot survive here. The decomposition of organic matter generates a large amount of acid that leads to developing podzolising soil types, of which there are podzolic, gray forest, illuvial-gley soils.

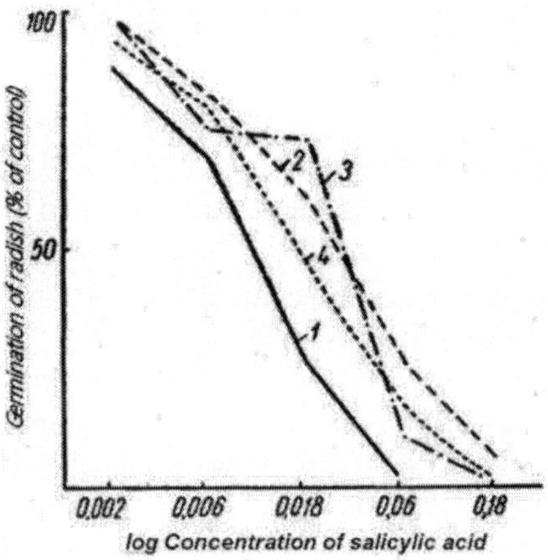

Figure 12. Effect of increasing concentrations of salicylic acid on germination of seeds of lupin (1), lettuce (2), radish (3) and corn (4) at K-50%.

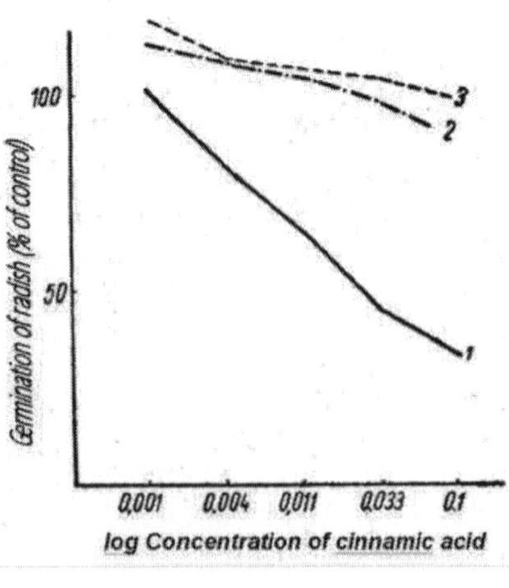

Figure 13. Effect of increasing concentrations of cinnamic acid on germination of seeds of lettuce (1), corn (2) and radish (3) at K-50%.

In grassy cenoses, many polyphenolic compounds enter through the litter, which is responsible for the accumulation of the bulk of the cholines in cenoses. Some researchers tend to explain generally all the chemical interaction of plants on the participation of polyphenolic compounds.

It is difficult to overestimate the allelopathic, and therefore the environmental role of tannins and polyphenols.

Their actual effects on higher plants are varied. Sometimes they are almost inactive. For example, a preparation of tannin very slightly delayed germination in bioassay in even high concentrations. In other cases, they can cause significant inhibition, which resemble that of coumarin.

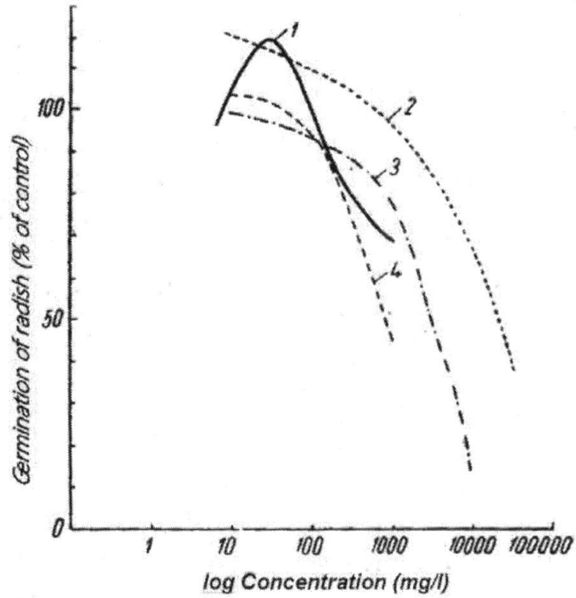

Figure 14. Influence of some polyphenolic acids on germination of radish at K-50%: 1 – Protocatechuic acid. 2 – tannic acid 3 – gallic acid 4 – cinnamic acid

For example, consider the effect of salicylic and cinnamic acids on seed germination. As can be seen from Figure 12, the salicylic acid was markedly toxic to the seed germination of all plants tested, being very sensitive to it over the entire range of concentrations as found with lupine and to a less extent - lettuce. On the contrary, cinnamic acid was not as active. In this case, the lettuce was depressed to a small extent, whereas the germination of radish and corn was markedly stimulated by weak solutions and little altered under the influence of high concentrations.

The curves in Figures 14 and 15 curves illustrate the different effects of phenolics and quinones respectively.

It can be assumed that these substances, that are very common in plant secretions, and that are quite resistant and little attractive to the bacteria, play a very important role in the allelopathic balance of phytocenosis. They are to a certain extent akin to humic acids as are their precursors. In cenoses, various intermediate of polyphenolic compounds are always present.

We are constantly determining the content of tannins in extracts and solutions that have been tested in bioassays.

The closest correlation between the content of tannins and inhibition of germination was observed in members of the family Rosaceae. Conversely, there was an inverse correlation in the Polygonaceae, while in the Cucurbitaceae, despite the exceptional strength of inhibitors, tannins are virtually absent. This does not exclude the possibility that in some cases inhibitors of germination are tannins, but we tend to think that more often inhibitors have a different nature.

The same applies to other types of plant exudates. In root excretions, tannins are almost absent, and guttational fluid, "honey-dew", etc. do not have them. Most tannin extracts are from leaves, bark, and stems. It is impossible to attribute all cases of inhibition just to tannins and phenolic compounds.

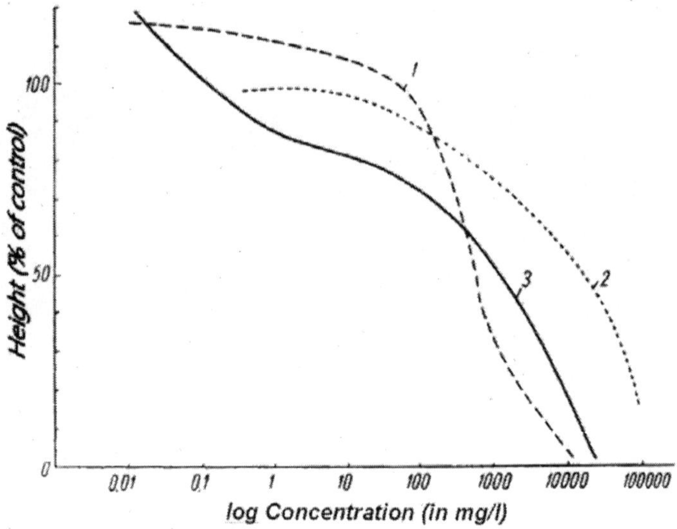

Figure 15. Effect of substances with quinone groups on germination of radish at K-50%: 1 – hydroquinone 2 – pyrogallol 3 - metol

In recent years, it has been shown that polyphenols participate in growth processes through interaction with endogenous auxins (Kefeli, 1963). The nature of this interaction has not yet been discovered. It is believed that some of the polyphenols may be precursors of heteroauxin[4]. They also alter the activity of the enzyme complex - auxin oxidase, which includes polyphenol oxidase.

The action of tannins on physiological processes in acceptor plants has been little studied. They can alter the permeability of the protoplast, denature proteins, and

[4] Indole acetic acid or IAA.

interact with metals. Probably, the absorption of tannins by plants is slow because they are large molecules and they reduce the permeability of the plasma membrane.

Anthocyanins and anthoxanthins. These are widespread water-soluble pigments in cell sap of plants. Sometimes they are in the cells in the crystalline state. Anthocyanins are derivatives of 2-phenylbenzopyrenes, and occur in plants, usually in the form of glucosides.

This group of pigments causes the variety of colours of flowers, fruits, stems, wood and other organs. By acid hydrolysis, pigments are broken down to anthocyanides.

Anthocyanins are functionally related in plants with flavonoids and tannins. They regulate the distribution of energy absorbed by the leaves, although they are not involved in photosynthesis. The presence of anthocyanins is apparently related to frost-resistance and durability against fungal diseases. We have repeatedly observed that allelopathically damaged plants acquire a bluish, purple, or pink colour, indicating the accumulation of anthocyanins. Tomato seedlings, that were planted after radish or tomato, were stunted, and contained 2-4 times more anthocyanins than the control plants,

In the literature there are indications of activity of anthocyanins and related compounds - flavones, flavonols, flavanones, which are pronounced inhibitors of germination or antibacterial substances.

The mechanism of their physiological action is similar to that of polyphenolic compounds.

Terpenes. The group of terpenes includes essential oils, carotenoids, resins and rubber, with differing chemical properties but grouped together based on common attributes, unsaturated bonds, volatility, lipid solubility, and primarily a carbon skeleton of similar structure. All of these substances are molecules composed of two or more isoprene units C_2H_5 combined in either open chains or in closed systems containing one or more rings.

The composition of the essential oils include aliphatic and alicyclic terpenes with 10 carbon atoms (monoterpenes) or 15 carbon atoms (sesquiterpenes), and their oxidation products. Essential oils and resins are found in special secretory cells or vesicles of some plant families, for example, in the resin ducts of softwoods, in the glandular cells and glands of some angiosperms (Labiatae, Rutaceae, Umbelliferae, Myrtaceae, and others).

In addition, unsaturated terpenes are found in nature as products of their oxidation or reduction – as alcohols, aldehydes and ketones. Essential oils have an effect on seed germination; in our experiments, we established their effect on the growth of roots, leaves and stems, guttation, photosynthesis, respiration and other processes in plants.

J.P. Cannon *et al.* (1962) have found the inability of Araucaria (*Araucaria cunninghamii* Ait.) to reproduce from seed under cover of *Backhousia angustifolia* Hook. *et* Harv., in secondary rain forests of Australia. The reason for this, according to them, is the inhibition of seed germination of Araucaria by the essential oil from the fallen leaves of *Backhousia*, which mainly includes angustione and dehydroangustione. These isolated substances in laboratory experiments delayed germination of Araucaria, cress and wheat, although the authors failed to show a toxic effect of dehydroangustione from whole leaves under natural conditions.

Very many researchers have established that the allelopathic effect of couch grass (*Elytrigia repens* (L.) P.B.) is associated with the release into the environment of agropyrene (Treibs, 1947; Laštuvka, 1955-1963; Grümmer, 1961; Aamisepp, Oswald, 1962):

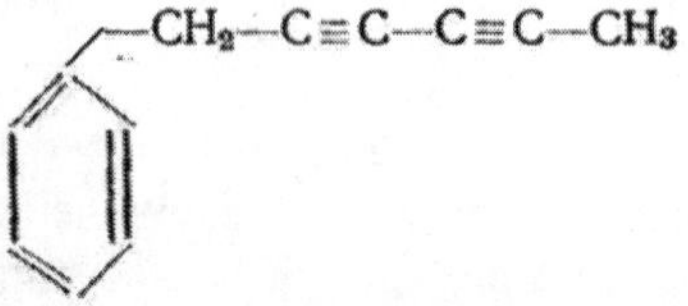

Agropyrene (1-phenyl-hexadene-2,4)

The essential oil distilled from the rhizomes of couch grass consists of 95% agropyrene. By dry weight, they contain 0.06% agropyrene. Agropyrene has fungistatic activity against fungi and bacteria, inhibits seed germination and the growth of the roots and stems of higher plants.

In long-term studies, Z. Laštuvka (1960) showed that agropyrene penetrates the roots or leaves of plants and is deposited in tissues in the form of brown masses poisoning various enzymatic processes in wheat, rye and other plants. Acting in the soil, agropyrene primarily affects root tips and limits the formation of root hairs. Because of this aggravation, the formation of secondary roots, is also soon terminated. This is then followed by the death of cells of the epidermis, cortex and even the central cylinder of the root. Agropyrene permeates the cell walls, fills the vessels and moves upward, poisoning the youngest parts of the plant.

Even with small concentrations, where under sterile conditions plants suffer no consequences, agropyrene acts detrimentally, as it reduces the resistance of wheat to the fungus *Stemphylium ilicis* Tengwall that settles on the roots and stems.

The essential oil of wormwood (*Artemisia absinthium* L.) contain β-caryophyllene, bisabolene and other compounds, forming in the air - chamazulene[5]. All these substances are able to inhibit the growth of roots, seed germination, the activity of oxidative enzymes (Schwär, 1962, 1963b). Similar essential oils have been isolated also from other species of wormwood.

Steviol is a diterpenoid, the aglucone of the glucoside stevioside from *Stevia rebaudiana* of the Compositae, which has gibberellic activity and causes the normal growth of dwarf mutants of maize (Ruddat, Lang, 1963). Perhaps steviol is a donor of structural units for the biosynthesis of gibberellins.

According to many authors, essential oils penetrate into the plant by diffusion through stomata and by dissolving in water and impregnating the cell walls. K. Höhn (1958), who studied the effect of essential oils on the roots (by guttation), believes

[5] Chamazulene is an interesting compound that is known foremost from chamomile, as in the course of distillation of the oil, chamazulene, a striking blue compound, is formed.

that they act like a drug, by covering the protoplast with a thin layer and preventing the absorption of oxygen and water.

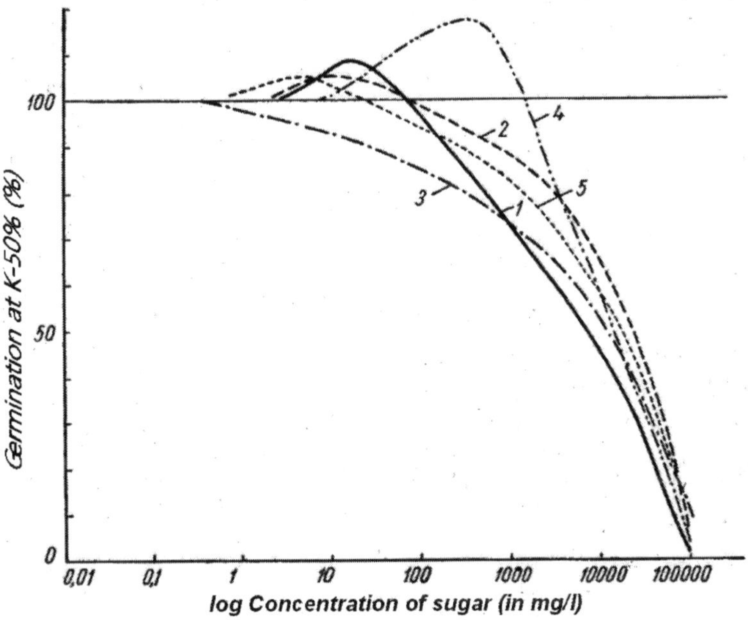

Figure 16. Effect of increasing concentrations of various sugars on germination of radish at K-50%:
1 – sucrose 2 – glucose 3 – mannitol 4 – lactose 5 - maltose

Essential oils are secreted in significant quantities. Thus, the air over a hectare of pine forest in hot weather can contain up to 30 kg of essential oils. The overall productivity of volatile terpenes in the world according to F.W. Went (1962) is 175 million tons annually. In his opinion, essential oils are the cause of electricity in the atmosphere, thunderstorms and tornadoes.

Carbohydrates. It is considered that sugar does not have toxicity towards higher plants. As found in some experiments, the retardation and inhibition of seed germination of lettuce is explained by the osmotic effect of sugars (Mohil, Paleg, 1963). Meanwhile, an iso-osmotic concentration of carbohydrates has a different effect on the germination of radish (Figure 16).

One of them, mannitol, has the most inhibitory effect, and stimulation of germination at low concentrations is absent. Other sugars stimulate germination at a dilution of about 10-50 mg/l, and with lactose – up to 500 mg/l. Retardation starts with various sugars according to different concentrations, though the maximal inhibition of germination for all tested sugars is at concentrations roughly above 5%.

However, the effect of sugar on germination is in the first place osmotic in nature, as evidenced by the curves of action of isotonic solutions of sodium chloride and sucrose. (Figure 17).

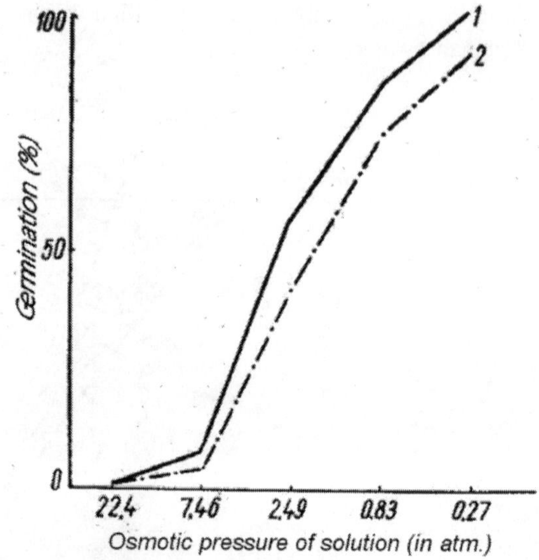

Figure 17. Effect of osmotic concentrations of sucrose (1) and sodium chloride (2) on germination of radish at K-50%.

In natural conditions, cenoses can rarely accumulate sugar in such amounts as to prevent seed germination and plant growth. There can be a real difference with sweet fruit or mass occurrence of "honeydew". However, small amounts of carbohydrates are everywhere. Thus, the soil, according to H. Deuel (1959), contains 10-100 mg carbohydrates per 100 g soil, from which more than eight simple sugars can be identified by paper chromatography.

Alcohols, aldehydes, ketones. The lower members of these groups of compounds are biologically very active volatile substances. They are able to be produced in higher plants, especially under adverse conditions of existence, and released into the environment. Many of these substances are formed during the ripening of fruits, when leaves are mechanically damaged (phytoncides), and with microbial degradation of organic matter under conditions of poor access to oxygen.

Ethanol is released by roots in anaerobic conditions (Grinev, 1961); it can accumulate in poorly aerated soil. Aldehydes and alcohols are released by leaves, and especially by maturing fruits and flowers.

The biochemical effect of alcohols is to shift some of the reactions in which they can participate directly, in blocking enzymes, and generally in poisoning the organism.

Acetaldehyde is contained in the immature caryopses of wheat and delays germination (Evenari, 1949). Benzaldehyde C_6H_5CHO is formed by hydrolysis of amygdalin and salicylaldehyde and is an extremely potent inhibitor of germination.

Growth substances. There is a specific interest to compare the natural cholines with synthesised physiologically active compounds. In addition, growth substances such as auxins, gibberellins and kinins are secreted by plants into the environment and have independent allelopathic value.

In the various organic products that get into the soil, J. H. Hamence (1945) found significant amounts of indole acetic acid (IAA).

IAA content in organic fertilisers in mg per 100 g dry matter was as follows:

Fresh manure	1.41
The same manure 4 months	0.15
Horse manure (not liquid)	0.80
Bird droppings (two weeks)	7.66
The same litter, dried	0.21 - 0.48
Bone meal	0.11
Dried blood	0.28 - 0.90
Horn meal	0.64
Fallen leaves (litter)	0.007

The high content of IAA in poultry manure causes obvious burns when too high doses are applied. It is probably necessary for the destruction of auxin as dictated by practical techniques - fermentation of the slurry or mash of bird droppings. Adding this kind of fertilisers increases the content of IAA in the soil up to a certain level, which can serve as a catalyst, or vice versa, as a depressive effect on plant growth. Burns observed on grass pastures, are also due to the high content of IAA in the urine of animals, rather than to the influence of osmotic concentration of salts.

I. Niggemann (1955) has discovered in peat IAA and like substances, that are apparently similar to estrogenic hormone. Regarding the content of IAA in the soil itself, opinions diverge. W. Flaig and H. Breyhan (1956) indicated that there was content of indole compounds in humic acids extracted from chernozem.

Vitamins. We have already noted that the higher plants release into the soil a variety of vitamins of groups B, and A, which serve as food for rhizosphere microorganisms. K. E. Ovcharov (1961, 1964) called attention to this aspect of allelopathy, and provided an overview of the work on this topic.

Currently, there is some detail that clarifies the possible formation of vitamins by bacteria and fungi. It was found that in principle they can synthesise all the necessary vitamins, but there are many microorganisms that are incapable of synthesis of a vitamin. In the studies of N. A. Krasil'nikov (1958), in 192 cultures of bacteria isolated from different soils of the Soviet Union, less than half of the cultures did not synthesise vitamin B_1. According to others (cited by Ovcharov, 1961), in 163 yeast cultures, 87% are in need of biotin, 35% - of thiamine and pantothenic acid, and 12% - of inositol.

Among the autotrophic algae in the world's oceans, about 50% are heterotrophic with respect to vitamin B_{12}.

On the other hand, some higher plants need a source of vitamins, at least at the beginning of their development. Thus, R. Bahme (1949) showed that fungi secrete vitamins that germinating seeds of orchids cannot synthesise. The addition of vitamin E to the culture medium was able to significantly accelerate the growth of seedlings of orchids. These facts indicate that the germinating seeds of orchids are not fully provided with various vitamins.

The roots of higher plants in isolated culture also are heterotropic with respect to certain vitamins that are necessary to be added to the culture medium. It is possible that the growth of entire root systems depends on the availability of vitamins in the soil and fertilisers. In particular, some authors explain the positive effect of organic fertilisers as due to the presence of vitamins (Manorik, Vasil'chenko, 1964).

Thus vitamins, despite small amounts of them in the environment of the phytocenosis, play a huge allelopathic role, providing mutual selection, mutual fitting of components in the cenosis

Among the products of metabolic and post-mortal decomposition of simple and complex proteins that we refer to as the second group of allelopathic active compounds, there are many physiologically active, nutritional or oligodynamic substances.

Proteins and polypeptides. In the secretions of plants are relatively low molecular weight proteins, which have enzyme function or are protective substances – phytoalexins, lysozymes. They are also active in higher plants.

A. Winter and H. Street (1963) isolated from the culture fluid in which roots of tomato had been grown for a period of 28 days a new natural auxin stimulating growth in various bioassays at a concentration of 0.1-1 mg/l and acting as an inhibitor at higher concentrations. After detailed study, it turned out to be a complex of 5-hydroxytryptophan with a peptide consisting of five amino acids located in the following manner: glutamic acid - arginine - alanine - valine - leucine. The accumulation of the peptide in the environment is one of the reasons for the retardation and cessation of the growth of isolated roots.

Similar indole-producing complex compounds, particularly IAA with amino acids or peptides have been found in many plants (Fransson, 1960, Kutaček, 1961, 1963).

The stimulating effect of skim cow milk used in the USA as a fertiliser is determined by proteins (Gander, 1960). Solutions with 0.03% animal protein also stimulated growth in bioassays. Higher concentrations slowed growth.

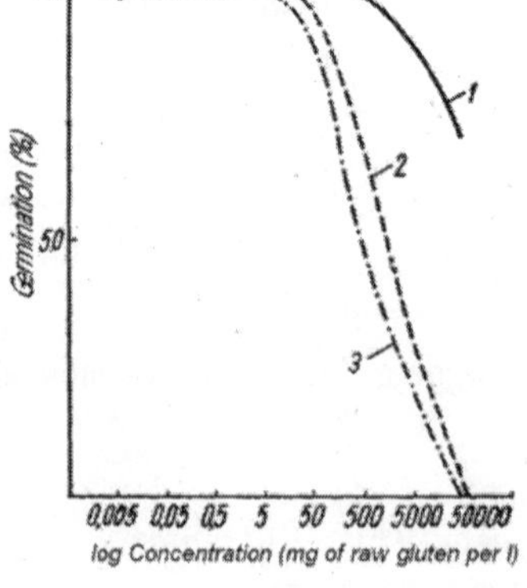

Figure 18. Effect of gluten hydrolysates from wheat grain on germination of radish at K-50%:
1 - water 2 – alkaline 3 - acidic

We have attempted to determine the effect of protein hydrolysates of wheat grain on the seed germination of radish. For this washed gluten was hydrolysed in 2.5% strength HCl, 2.5% strength NaOH 2.5% in water. Gradually the diluted hydrolysate was neutralised to pH 5.4 and was tested in bioassay. The results are shown in Figure 18.

It can be seen that protein degradation products have a very high toxicity and cause complete destruction of shoots at high concentration. Stimulation of germination was found.

Derivatives of nucleic acids - nucleotides, nitrogenous bases, and others can have very high physiological activity. According to Z. Šormovo (1962), they seem like typical growth regulators. Some derivatives, such as 5-bromouracil, are strong chemical mutagens.

In recent years, much attention has been attracted by kinetin or 6-furfurylmethylaminopyrine:

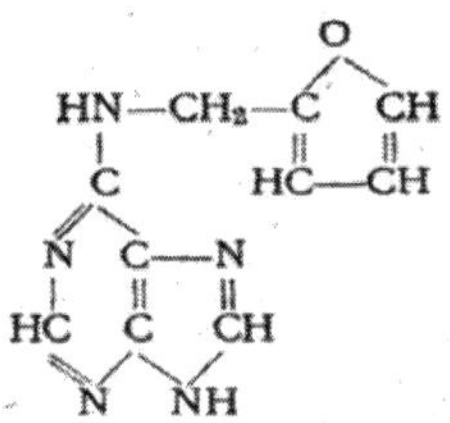

It was found initially in yeast, then in tobacco and other plants. Kinetin is formed from the breakdown of nucleic acids[6] and it has such an unusual action that it is in a separate group called kinins. Kinetin often acts as an antagonist of IAA, and removing it causes growth inhibition. Unlike IAA, it acts primarily on division of the cell and to a lesser extent on cell extensibility; it accelerates protein synthesis, increases the stability of the chlorophyll-protein complex, and it slows down the ageing of organs. When applied topically, kinetin causes increased flow of nutrients to the treated organ.

Obviously, in nature these can often occur as accumulations of proteins undergoing decomposition. Therefore, the value of allelopathic substances described here is very large.

Amino acids. As noted above, the amino acids are part of the components of plant secretions made in its lifetime. Large quantities of them are formed during the decomposition of plant materials.

The high physiological activity of some amino acids has long been known. Thus, tryptophan is a precursor of IAA, to which it easily is converted, and it takes indirect participation in the processes of respiration. In this regard, it retards the growth of the roots of lentil (*Lens culinaris* Med.) at a concentration of 10^{-7} to 10^{-4} M

---

6 Kinetin was initially thought to be a breakdown product, but is now known to be synthesised in plants.

and is stimulatory at weaker concentrations (Pilet, Kobr, 1958). The concentration of endogenous auxin in tissues is increased.

A.M. Smirnov (1961) and E.I. Ratner and A.M. Smirnov (1963) showed that some pure preparations of amino acids are often toxic to the growth of isolated roots. Valine proved to be an inhibitor of the growth of isolated roots already even at a concentration of 3 mg/l. Glycine, tryptophan, asparagine, cystine stimulated root growth at 10-25 mg /l with a background of nitrate nitrogen, but at higher concentrations, they inhibited growth. Individual amino acids that accumulate in tissues, that are poorly transformed, have a negative affect on the roots. In the case where a mixture of two or more amino acids is administered, the toxic effects are considerably mitigated or give way to stimulation, as plants more fully use the amino acids as fertiliser

The effect of some amino acids on seed germination is shown in Figure 19.

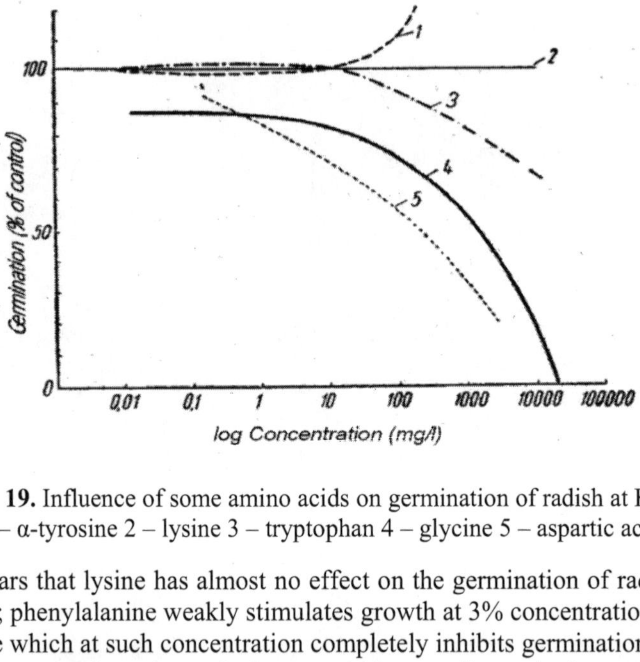

Figure 19. Influence of some amino acids on germination of radish at K-50%: 1 – α-tyrosine 2 – lysine 3 – tryptophan 4 – glycine 5 – aspartic acid

It appears that lysine has almost no effect on the germination of radish, lettuce, lupine, corn; phenylalanine weakly stimulates growth at 3% concentration (compared with sucrose which at such concentration completely inhibits germination!). Tyrosine stimulated germination, but glycine, ornithine, valine, alanine, aspartic acid dramatically delayed germination. Tryptophan in high concentrations causes average retardation, and stimulation is not observed, which suggests a similar action of tryptophan and IAA, which in these bioassays does not cause stimulation.

A.V. Blagoveshchenskii (1951, 1953) and E.V.Kolobkova and N.A. Kudryatsova (1956) found that the inhibitors from the seeds of the Siberian peashrub is a monoaminomonocarbonic acid, and suggested that other inhibitors may be naturally occurring amino acids. It should be borne in mind that in addition to the usual 20 constituent amino acids that make up proteins, about 100 different amino acids are found in plants.

During the study of the inhibitors of germination in the fruit and seeds using bioassays, we have simultaneously determined the total content of the ninhydrin positive compounds. Correlations found for relationships from 730 samples are shown in Table. 19.

From Table 19 it can be concluded that the presence of amino acids and similar compounds in the extracts is more closely correlated with their inhibitory activity than was found for tannins.

Data with a correlation coefficient above 0.6 accounted for 68.6%[7]. The weighted average of correlation coefficients for the total was 73.6%. In this analysis, the number of cases of positive deviations was greater than the number of negative by a factor of 2.3, i.e. not as much as with the tannins.

Thus, it should be recognised that the supposition by A. V. Blagoveshchenskii, that there is an amino acid nature of many inhibitors of germination, is confirmed to some extent in these experiments.

Comparing the correlation coefficients for members of individual families, it is necessary to emphasise the high values in the Ranunculaceae, Leguminosae, Compositae and Rubiaceae and low correlations in Euphorbiaceae, Aceraceae, Gramineae, Umbelliferae, which, obviously, is due to inhibitory compounds being of different chemical nature.

Table 19. Correlation among the content of amino compounds and the inhibition of germination

Coefficients of correlation	Correlations as percentage of the all samples		
	Positive	Negative	Total
1.0	–	–	27.6
0.8	22.2	11.6	33.8
0.6	11.4	5.8	17.2
0.4	11.0	3.2	14.2
0.2	4.7	0.8	5.5
0.0	1.1	0.6	1.7
Total	50.4	22.0	100.0

Amino acids are commonly found in nature. F. Scheffer (1958) provides an overview of the work, which shows that almost all of the amino acids that are found in plants are found in the soils. K. Dadd *et al.* (1953) found that the solution, pressed from the soil had up to 5 mg of nitrogen as amino acids per 1 ml. In acid soils, the content of free amino acids is relatively higher, and in the spring it is higher than in summer. This is due to a greater decomposition as the conditions are optimal for the growth of microorganisms.

The accumulation of amino acids in the soil during the winter can be explained by their release from crop residues and roots. On the other hand, it was found that growing plants excrete significant amounts of amino acids (Frenzel, 1957;

[7] It is not apparent how this figure was derived.

Katznelson *et al.*, 1955; Linskens, Knapp, 1955; Parkinson, 1955; Pearson, Parkinson, 1960-61; Scharpenseel , Krausse, 1962).

Amino acids in the soil are used fairly quickly by the microflora (Rovira, 1956), so it is difficult to expect a significant direct effect on the neighboring plants.

Given the undoubted influence of amino acids on the formation of specific microflora one can expect that there is also an indirect allelopathic effect due to this part of plant excretion.

There is nothing known concerning the presence of other amino derivatives in the environment of phytocenoses - amines, amino alcohols, amino sugars, etc. These substances, often with very high activity, are products of the hydrolysis of proteins, amino acids, alkaloids, phosphatides, and other compounds, but can also occur as the result of transamination.

Ammonia. Of no small allelopathic importance is the end product of nitrogen metabolism in plants - ammonia, which is known to have toxic effects on living organisms.

Some authors consider ammonia and ammonium ions as inhibitors of germination, which are secreted in the environment due to swelling of the seed balls of sugar beet (Stout *et al.*, 1941a, 1941b). It was shown that there was a correlation between the content of NH_4^+ in extracts of sugar beet seed and growth inhibition. However, we now know that the inhibitors in beet seeds have a more complex nature.

G. M. Vines and R. T. Wedding (1960) believe that ammonia non-specifically alters the permeability of the protoplast and also suppresses respiration associated with oxidation of diphosphopyridine nucleotide (DPNH). Ammonia salt solutions were more active than gaseous ammonia at the same molar concentration; ammonium sulphate or chloride is more toxic than ammonium phosphate or carbonate.

Our data show that ammonium salts markedly inhibit seed germination of radish, and the effect is determined by the accompanying anions. In particular, nitrates and phosphates enhance the effect of ammonia at low concentrations, as do oxalates and phosphates at higher concentrations. At concentrations of ammonium ions lower than 50-100 mg/l, toxicity is virtually absent.

In extracts from the soil litter we are finding ammonium ions, the content of which ranged from 0.1-0.2 to 7-16 mg/l. Such non-toxic concentration of ammonia is essential to ensure the nitrogen nutrition of plants. However, it is possible that in some cases (for example, under anaerobic conditions), the ammonium ion content can reach toxic values.

Gaseous ammonia has been observed by us (Grodzinskii *et al.*, 1962), occurring as a constant component of miasmins – volatile secretions from decomposing plant mass.

There is a third group of compounds having great allelopathic significance – these are the exudates of heterotrophic organisms that are specific (antibiotics, marasmins, gibberellins, vitamins, etc.), and nonspecific, that form the mobile mass of organic substances, which is called soluble humus.

Consider some specific substances produced by lower heterotrophic organisms.

Antibiotics. Some researchers have imbued this term with a very broad meaning, referring to antibiotics always as those which adversely affect any living organisms (Duquenois, 1958), but more often with this term, one understands the substances produced by fungi or actinomycetes, and that have antibacterial action.

Table. 20. Characteristics of physiological effects of some antibiotics

Name of drug	Producing organism	Effect on plants
Aureomycin	Streptomyces aureofaciens	Inhibits growth of virus particles. At 10^{-8}-10^{-6} M stimulates, at 10^{-4} M inhibits growth of isolated tissues
Biomycin	*Actinomyces aureofaciens*	Delays germination, disrupts biosynthesis of chlorophyll in seedlings
Gramicidin C ($C_{30}H_{46}O_5N_6$)	*Bacillus brevis*	Affects Gram-positive and Gram-negative bacteria
Griseofulvin	*Actinomyces griseus*	Poisons plants at 5 mg/l; inhibits growths; causes thickening of roots
Levomycetin (chloramphenicol, chloromycetin) ($C_{11}H_{12}O_3Cl2$	*Actinomyces venezuelae*	Specific inhibitor of protein synthesis in plants
Penicillin ($C_9H_{11}O_4N_2S$—R)	*Penicillium notatum, P. chrysogenum, Aspergillus sp.*	Active against Gram-positive bacteria. For higher plants indifferent
Synthomycin	Synthetic drug derived from levomycetin	Effects similar to levomycetin, but 50% less
Streptomycin ($C_{21}H_{39}O_{12}N_7$)	*Actinomyces globisporus, Streptomyces griseus, S. bikiniensis*	Inhibits growth of roots, stems and leaves; inhibits biosynthesis of chlorophyll, causing chlorosis. Seed germination not affected
Terramycin ($C_{22}H24O_9N_2$)	*Actnomyces rimosus*	Inhibits growth of primary root, resulting in general plant poisoning., No effect on germination

The effect of antibiotics on higher plants have been relatively little tested. Table 20 gives an overview of the known data on the subject.

As shown by many authors (Scheffer, Kloke, 1954; Gray, 1955; Visona, Pesce, 1963), streptomycin causes acute poisoning of plants, chlorosis, stunting of leaves, death of roots, inhibition of chlorophyll biosynthesis. Spraying manganese salts (Gray, 1956) partly removes the negative effect of streptomycin. Chelation of streptomycin also changes the nature of its physiological effects.

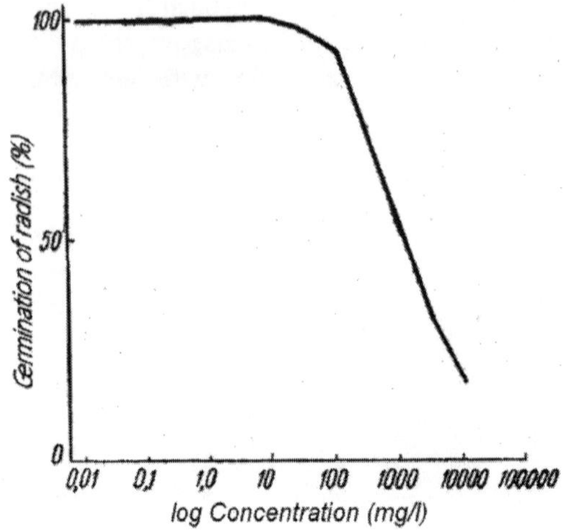

Figure 20. Effect of levomycetin on germination of radish at K-50%.

Antibiotics physiologically interact with endogenous auxins, or those found in the environment. For example, tetracycline, oxytetracycline and chlortetracycline in the presence of IAA stimulated the growth of isolated tissues of Jerusalem artichoke at concentrations of 10^{-8} - 10^{-7} M, and depressed growth at 10^{-4} M. Without IAA, stimulation did not occur (Netien, Lacharme, 1960).

Among the antibiotics that have a very strong effect on higher plants should be noted chloramphenicol, or levomycetin which is a specific inhibitor of protein synthesis.

In bioassay with radish, chloramphenicol was not particularly active. The stimulating effect of low concentrations did not appear; inhibition began at a concentration of about 100 mg/l chloramphenicol. The maximum concentration tested by us was 12670 mg/l antibiotic, which was not lethal for germinating seeds of radish. The corresponding concentration curve is shown in Figure 20.

This can be explained obviously by the fact that the initial phases of germination of seeds have little to do with the synthesis of new protein and occurs mainly due to swelling and expansion of the cells.

Chloramphenicol, is believed to not affect the respiratory cycle enzymes at low concentrations. At high concentrations of the order of 8000-16000 mg/l chloramphenicol inhibits respiration and the activity of α-amylase, and partially β-amylase, and it enhances the activity of proteinase in barley (Srivastava *et al.*, 1962). It blocks the formation of chlorophyll and the greening of seedlings (Molotkovskii, 1963).

Marasmins and gibberellins. These substances are secreted by many phytopathogenic and saprophytic microorganisms, or develop in the tissues of higher plants under the influence of pathogens. Very often, the anatomical and physiological changes in the plant, are caused by active substances, where the expedient use of such substances has arisen during a long evolutionary adaptation of the metabolism of the plant host, and parasites or semi-parasites.

Ultimately, regardless of the site of formation, marasmins and gibberellins play an important role in the relationship between higher plants and heterotrophic organisms.

Some saprophytic microorganisms that populate the surface area of roots and rhizosphere of plants, can seriously affect the growth and health status of plants through their metabolites. Often in their role of saprophytes, they act as typical parasitic forms, which after external toxic weakening of the plants, attack them. Adverse conditions, such as drought, contribute to this transition of the fungus from saprophytism to parasitism. This has been shown, for example by Ya. Lasik and M. Stanek (1963), for the parasitic bacteria *Xanthomonas fuscans* (Burkholder) Burk., which affects beans. The bacteria occur on the surface of the seed, and after germination colonise the root surface. In an environment with bean root exudates, the bacteria produces a toxic substance that inhibits seed germination, growth of the root system and above-ground parts of the plant.

In other cases, plant poisoning can occur under the influence of toxic metabolites from normal saprophytic fungi - *Fusarium*, *Pythium* and others. L.A. Nezgovorov (1959, 1963) found that the death of heat-loving plants (corn, cucumber, melon, pepper) when cooled to low positive temperatures 2-6° C, was associated with increased development of fungi, that poisoned the plants with their secretions. Obviously, a similar phenomenon causes the disease of sugar beet seedlings, "Black Leg", that occurs in waterlogged, cool soils with poor aeration. Among the agents of this disease include more than seventy species of saprophytic fungi and bacteria which attack weakened plants and cause their ultimate death. Seed treatment with antifungal drugs increases the cold resistance of plants and resistance to Black Leg.

From the culture filtrate of *Aspergillus niger* there have been isolated malformins (Takahashi, Curtis, 1961), which are neutral cyclic peptides consisting of four amino acids - valine, leucine, isoleucine and ½ of cystine, with the general formula $C_{23}H_{39}N_5O_5S_2$. Different strains of the fungus produce several different varieties of malformins. For example, instead of isoleucine in the molecule there can be allo-isoleucine.

According to R.W. Curtis (1961), malformin already has a significant effect at a level of 0.0000001 mg per plant; the strongest effect was found at a dose of 0.0001 mg per plant. It is absorbed by the roots from the soil and moves in the aerial parts. Dry or wet weight of the plants does not change significantly, but malformin causes a curvature of roots and stems, and various morphological deformities. Malformin has a broad spectrum of action and damages many dicotyledonous plants; monocots are more hardy.

Unfortunately, the chemical nature of marasmins, malformins inhibitors and stimulators has been studied very little. Also poorly studied are their physiological and biochemical effects on higher plants. An overview of some known substances of this type is given in Table 21.

Table 21. Physiological activity of substances, produced by lower plants

Substance	Chemical composition	Producer organism	Physiological effect
Alternaric acid	Unsaturated, diacid $C_{21}H_{30}O18$	*Alternaria solani*	Phytotoxicity, specifically to plants of Solanaceae
Vasinfuscarin	Enzymic protein	*Fusarium lycopersici* *F. vasinfectum* *F. moniliforma* *Botrytis cinerea* *Fusarium solani*	Darkening and blockage of vessels, wilting
Botriticin	—	*Verticillium alboatrum*	Inhibits alcoholic fermentation
Diaporthin	$C_{13}H_{14}O_5$	*Endothia parasitica*	Necrosis of vascular bundles, wilting
Ipomeamorone	Sesquiterpenoid	*Ceratostomella fimbriata* (on potato)	Protective substances formed in affected tissue, causes uncoupling of respiration and oxidative phosphorylation
Itaconic acid	—	*Helicobasidium mompa* Tanaka	Accelerates decomposition of pectin
Lactones, α-lactilamino-β-hydroxy-*l*-aminopimelic acid	Lactone	*Pseudomonas tabaci*	Necrotic spots on leaves of tobacco, antimetabolite of methionine, donor of methyl groups
Lysergic acid	—	*Claviceps purpurea*	Part of the ergot alkaloids
Lycomarasmin	Dipeptide	*Fusarium Lycopersici*	Phytotoxins - forms necrosis on peripheral parts of tomato leaves. Forms chelate complexes with iron and copper
Pericularin	$C_{17}H_{14}C_2O_3$ Unknown structure	*Pericularia oryzae* Cavara	Binds chlorogenic acid
α-picolinic acid	C_7H_5O2	*Pericularia oryzae* Cavara	Inactivates porphyrin-containing enzymes, forms chelates
Skyrin	Bianthraquinone $C_{30}H_{18}O10$	*Endothia parasitica*	Necrosis of vascular bundles, wilting
Phytonivein	Sterin, $C_{29}H_{46}O_2$	*Fusarium oxysporium f. niveum*	Inhibits germination of seeds; wilting of plants
Fusaric acid	5-butylpicolinic acid 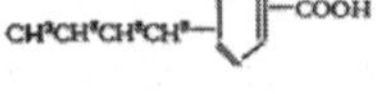	*Fusarium heterosporum* *F. lycopersici* *Gibberella fujikuroi*	Strong osmotic effect on protoplast; increases permeability; increases respiration through enzymes containing porphyrin; forms chelates with iron. Causes necroses between veins of tomato leaves.

Various indole products, vitamins, enzymes, etc., have long been found among the excretions of microorganisms and fungi

Among the most studied substances of fungal origin are the gibberellins, which are the focus of a number of experimental, studies and several reports (Stove, Jamaki, 1957; Khailakhyan, 1961; Muromtsev, Pen'kov, 1962).

Gibberellins were initially detected in secretions of the fungus *Gibberella fujikuroi*, for which the conidial stage is called *Fusarium moniliforme*; and since then, these substances have been found in many plants. They are potent activators of growth, but in contrast to IAA at high concentrations, gibberellins are nontoxic. Gibberellins strongly stimulate the growth of stems via elongation, cause long-day plants to bolt from their short-day rosettes, and can return genetically dwarf forms of plants to their original size. Treatment with gibberellins often leads to chlorosis, abnormal elongation of stems, and generally it has an impact on growth that is opposite that of red light. It is believed that gibberellic acid blocks the activity of the groups of enzymes oxidising IAA (auxin oxidase, peroxidase, flavin oxidase, etc.), which results in an accumulation of endogenous growth substances, and growth enhancement.

Gibberellins alter the overall direction of the plant metabolism such that there is reduced synthesis of polymeric compounds - cellulose, starch, and nucleic acids, etc., so that there is a relative enrichment of mobile forms of carbohydrates and other substances.

Currently, there are about ten gibberellin substances. Essentially they are active when the molecule contains a ring.

K. M. Sytnik[8] discovered something important for allelopathy, the fact that in pumpkin plants growing in the same container, gibberellins move from one plant to another. The same phenomenon for dwarf pea was observed by R. Knapp (1960b).

Gibberellins act on almost all plants; a few exceptions can be explained by imperfect technique of administrating the substance inside the plant. Gibberellins accelerate the germination of seeds, and they can replace the action of light for the seeds that do not germinate in the dark.

There are many works that show the physiological effects of metabolites of bacteria and fungi with unexplored chemical structure. The study of the microflora of the rhizosphere of different plants in different climatic zones allows the determination of a large number of active species and strains of soil fungi and bacteria. The secretions of these microorganisms have a very strong influence on the growth of higher plants. In recent years, much attention has been placed on the effect of soil micromycetes on various crop plants. Most often examined were the effects of culture fluids at various dilutions, included in the nutrient solution in which the test plants were grown, or deposited on leaves, etc. Among these studies, should be noted those by V. Bilai (1952, 1955), who in 1949-1953, studied the effect of species of the

---

[8] K.M. Sytnik is omitted from Grodzinskii's bibliography, but one can assume that the article is the same as that cited by Grodzinskii in his 1973 work *Osnovy Khimichnoi Vzaemodii Roslin*, and is given in the bibliography in this translation.

genus *Fusarium* on seedling growth of wheat, corn, rye, barley and cotton. She noted a positive effect on plants by excretions of *Fusarium avenaceum*, *F. semitectum*, *F. moniliforme*.

V.S. Moskovets (1961) found that *Aspergillus candidus*, *A. oryzae*, *P. rugulosum*, *Fusarium avenaceum*, *F. merismoides*, *Actinomucor corymbosus* stimulate the growth of alfalfa seedlings. In studying the antibiotic properties *Penicillium* species (section *Monoverticillata*), S. M. Kharchenko (1961) observed stimulation of cabbage seedlings by *P. bilaii*, *P. multicolor* and *P. terlikowskii*. T.G. Mirchink and K.P. Greshnikh (1961) showed that the most toxic species to plants are *P. cyclopium*, *P. purpurogenum* and *P. martensii* isolated from soddy-podzolic soils from the Moskow and Voronezh regions and from the alpine soils of Pamir.

In studying the harm of soil mulching with straw and crop residues, which is observed in wet years in Nebraska, D.E. Behmer and T.M. McCalla (1963) attributed this to the influence of metabolic products of fungi that decompose organic matter. In their experiments, the addition to the soil with 1% by weight of the milled residues of alfalfa, corn and sorghum, significantly slowed the growth of wheat. When soil was inoculated with *Penicillium urticae* Bainer, the inhibitory effect was enhanced; in soil treated with the residues of alfalfa, roots were observed to have curvature, similar to those caused by malformin. One of the toxins of *P. urticae* is patulin.

N.M. Podoplichko *et al.* (1963) have isolated fungi from the rhizosphere of corn, and have studied the effect of the film and culture fluid on corn seedlings. They tested 84 species of fungi. The authors note that diluted solutions contribute to a better determination of the inhibitory or stimulating effect. Among the species tested that had the most stimulating effect were *Penicillium lanosum*, *P. bilaii. Cladosporium griseo-olivaceum*, *Cl. linicola*, *Cl. transchelii*, *Spicaria violacea*. There were many kinds of different effects on the growth of roots and above-ground parts. For example, *Fusarium javanicum*, *F. avenaceum*, *P. frequentus*, *Cl. transchelii*, *Cl. linicola* stimulated root growth, and *P. lanoso- coeruleum*, *Congronella butleri*, *P. thornii*, *P. spinulosum*, *A. terreus*, and others stimulated growth of the aerial part of the corn seedlings. Stimulation is best detected when the filtrate is diluted into 100-100,000 times.

There was great retardation of growth of seedlings with *P. rubrum*, *P. duclauxi*, *P. expansum*, *Cl. brevi-compactum* v. *tabacum*, etc.

It may be noted that the effect of culture filtrates on higher plants is nonspecific. As a rule, low concentrations of metabolites cause stimulation of growth, and more concentrated solutions cause suppression and death of plants. The metabolites of various microorganisms act in a fairly similar manner to those of various types in higher plants. All this testifies to the complex composition of secretions, which physiologically interact with a wide variety of compounds. It is also obvious that the higher plants are not indifferent to metabolites constantly accompanying bacteria and fungi. These metabolites play a role in the life of higher plants, affecting all aspects of metabolism, growth and production, resistance to diseases and pests, etc. They play a very important role in indirect allelopathy, in the selection of a harmonious, coherent composition of the phytocenosis etc. The study of these substances is of great interest.

Even of more nonspecific nature are the physiological effects that are attributed to the products of more profound microbiological transformations of organic matter in the soil. Thus, as is known, there occurs a mixture of more or less stable organic products, the so-called humic acids or humus.

It can now be considered firmly established that humic substances are not individual compounds with a defined chemical structure, but are a mixture of different polymers containing heterocyclic, phenolic and quinone groups. In total, they represent 85-90% of the organic matter in the soil (Kononova, 1963).

It is generally known that there is a beneficial effect from humates extracted from soil for various physiological processes in plants (Kononova, 1963; Khristeva, 1948, 1951, 1955; Kloke, 1963). Preparations of soluble humates made from humus, peat, manure in sand and soil cultures, in a number of plants elevated the yield by 25-46%, increased the content in tissues and uptake of nitrogen, phosphorus and potassium (Khristeva *et al.*, 1954; Ronsal', 1963) . Humic fertilisers accelerate the flow of P^{32}, while at the same it increase the sugar content, chlorophyll, and peroxidase activity. Infiltration of humates into leaves also enhances phosphorus flow, photosynthesis and respiration. It is believed that humic substances can act as dehydrogenases, as hydrogen acceptors in respiration (Guminski *et al.*, 1955). P. Kaul (1962) believes that the polyphenols from decomposing organic material in the soil, penetrating into the plants, can accelerate the transfer of electrons and hydrogen, and thereby enhance the growth, in particular, as they specifically activate a copper terminal oxidase.

In general terms, the effect of humates from humus does not differ from that of humate obtained from peat, lignite or soil humus (Ronsal', 1963).

The problem of humic substances, humic fertilisers, natural stimulants such as "petroleum growth substances" is close to the problem of allelopathy to the extent that these stimulants have a role in cenoses and a mechanism of action similar to the action of the complex mixtures of allelopathic active substances which are released by plants and microorganisms.

In conclusion, it is necessary to make a general remark: almost never with pure synthetic drugs can one get such a clear inhibition, as with natural extracts, if they are taken, of course, in comparable concentrations. Natural extracts are much more active.

Sources of resistance and destruction of cholines in phytocenoeses

The chemical nature of cholines participating in the cycling of active substances in phytocenosis, in most cases, remains unknown. What is clear is that the primary sources of these substances are green photosynthetic plants that create a certain mass of organic matter per unit area. In the end, sooner or later, all of its organic material dies, is expropriated from plants in the environment where it is more or less completely decomposed by heterotrophic organisms, that mineralise them to the simplest components; only a relatively small portion of the organic matter is converted into permanent humus. In the pathways of all of these transformations, part of the organic matter exists in a physiologically active form, capable of providing an oligodynamic effect on higher plants and other organisms. At the beginning of these pathways, there are active substances formed by the plants, but in the course of further transformations they increasingly lose this specificity, become equal, acquiring properties of the complex, that are characteristic of the soil.

The general character of the cycle of organic matter in the soil can be represented in the scheme by A. Kloke (1963) given in Figure 21.

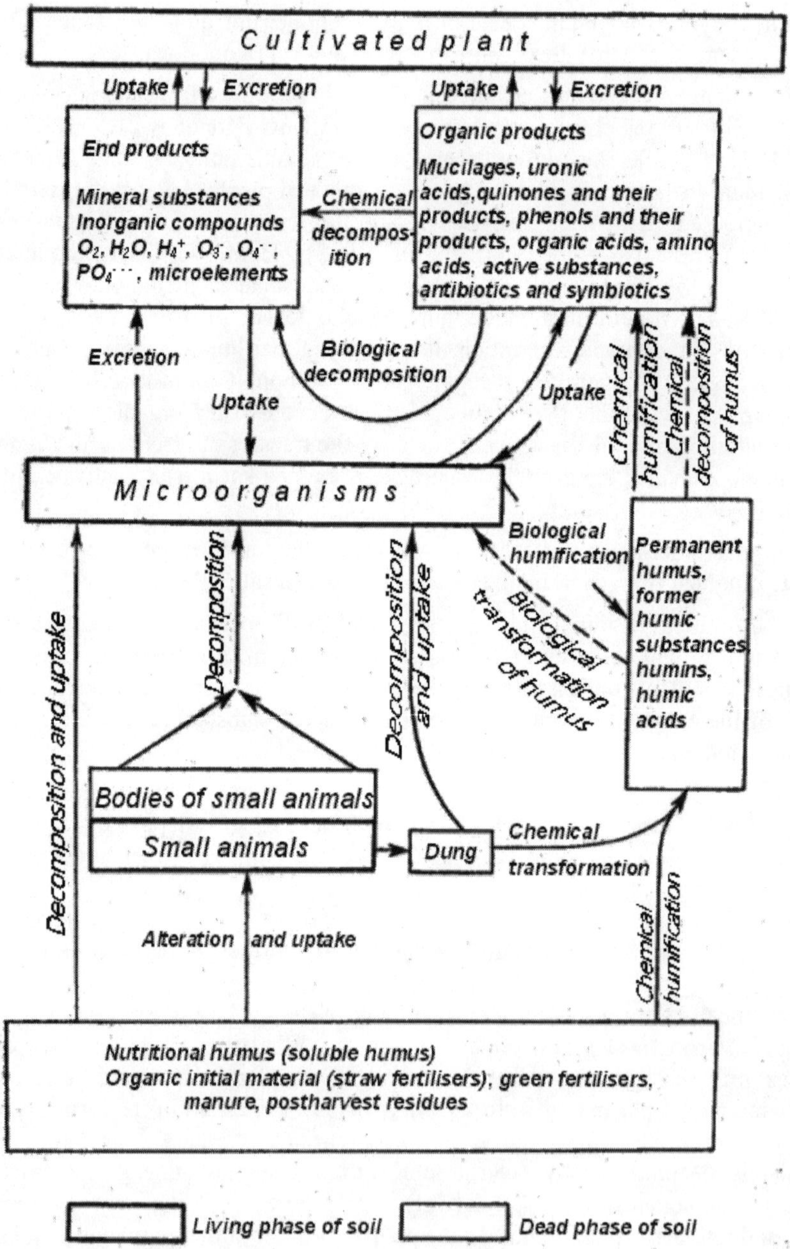

Figure 21. Circulation of organic substances in soil according to A. Kloke.

The accumulation of physiologically active products in the environment (mostly soil) is due to the combined action of a number of factors, the most important of which are summarised in Table 22.

Table. 22. Parameters, influences on the accumulation and fate of physiologically active cholines in the environment (soil).

Parameter	Degree of development of parameter, increased toxicity to acceptor plant	Degree of development of parameter, reduced toxicity and promoting favourable effects
Total productivity of their biomass in the cenosis	High	Low
Mass and rate of release of active substances by the donor	High	Small
Toxicity and persistence of substances	High	Low
Rate of diffusion in the environment	Average	Very high or very low
Adsorptivity of the environment	Average	Very high or very low
Leaching (washing) in the environment	Poor	Strong
Degradation by microorganisms	Poor	Strong
Interaction with other substances	In the nature of synergism or additivity	In the nature of antagonism
Interaction with other factors	Increases effect of cholines	Reduces effect of cholines

To assess the overall allelopathic level in the cenosis, there is the overall production of organic matter, the nature of its dying off, and decay. It can be assumed that the rate of decomposition of cholines is proportional to the speed of total mineralisation of organic matter. Some idea of this may be offered in Table 23, which is compiled from data which we have taken from the works of E. M. Lavrenko *et al.* (1955), L. E. Rodina (1963), I. Dommergues, 1963), and others.

The annual quantity of litter in the tropical zone, accordingly, is 2-5 times greater than in the temperate zone; the rate of decomposition of organic matter and the scale of cycling of nutritional elements is also considerably greater than in the north. However, the role of root exudates and root litter is difficult to consider, but it is likely that they obey the same laws.

Table. 23. Accumulation of biomass in various types of vegetation (in centners/ha)[9]

Type of vegetation	Total biomass	Annual growth
Arctic tundra	50	12-25
Taiga, conifer forests	1000-3000	45-85
Broadleaved forests of the temperate zone and subtropics	4000-5000	100-130
Savanna of the semi-humid regions	300-600	25-50
Temperate arid steppes, subshrubby deserts	70-700	25-115
Subshrubby deserts	40-50	12-25
Tropical forest of the humid zone	10000-17000	800

In northern well-watered regions, there are relatively more developed aerial organs of plants and their ratio to the roots reaches 2:1 – 1:1 and more. In the southern arid areas it is mainly the root parts that develop, so in some desert plants, the ratio is close to 1:25. The roots of the total stock of organic matter in a cenosis accounts for (in %):

Forest communities:	22 – 25
Steppe:	80
Subshrub desert:	85
Saxauls:	90

In parallel with these changes, there is the relative role of litter produced from aboveground debris and root residues in the soil.

J. Weaver (1962) suggests that the roots of herbaceous steppe plants live up to 12-15 and even 25 years, while the above-ground parts of the plant die annually.

Thus, in the cenosis there is annually tonnes of organic matter that are produced and subjected to various changes. What proportion of it is physiologically active, what role does it play in the existence of certain cenoses, what are the dynamics of this matter, etc., we do not know, as such studies are lacking.

It can be assumed that in the conditions of the Ukraine, the annual production of dry organic matter derived one way or another from the plant varies from 3 to 30 tonnes per hectare. Taking the weight of topsoil per hectare as 3000 tonnes, it can be calculated that the amount of organic matter falling to the ground or into the soil for the year ranges from 0.1 - 0.3%. The distribution of organic residues in the soil is uneven; there is a higher concentration of residues in the surface layer. Only later, as a result of decomposition and diffusion of soluble substances, and thus the activity of animals, is the organic matter mixed with the whole soil. This creates conditions for the uneven distribution of active substances among the structural elements for the cenosis, for the occurrence of certain areas of more active zones, or foci located among the less active medium.

Apart from the overall and the maximal quantity of cholines formed in the cenosis, it is of considerable importance that time has passed since the beginning. So,

[9] A centner, as used in eastern Europe, is a non-SI unit of 100 kg.

as noted in the works of R. Bode (1958) and in our multiple determinations of an autumn maximum of excretion of large amounts of cholines from dead leaves, it does not have great phytocenotic significance, because the plants at this time stopped their vital activity and were insensitive to the active substances.

The most important site in the environment of the phytocenosis for the accumulation of cholines is the soil, because the atmosphere is subject to rapid change, the aqueous medium of the ground of the cenosis is essentially inseparable from the soil. The soil generally has some toxicity to various organisms. The reasons for this phenomenon are manifold. On the one hand, it can be caused by a lack of nutritional substances, an unbalanced salt composition, etc. On the other hand, there may occur in the soil an accumulation of various toxic products that are formed as a result of anaerobiosis.

Biological processes in flooded, compacted and excessively wet soils, leading to reduction of nitrate to ammonia, sulphates to hydrogen sulphide, facilitate the transition of iron and manganese to the poisonous ferrous form[10]. Phosphoric acid forms a toxic salt of ferrous iron - blue vivianite. Carbon is cleaved from organic molecules in the form of methane and other hydrocarbons. All of these compounds are toxic, even to plants adapted temporarily for anaerobiosis.

In Japan, about 3 million hectares, or half the arable land is used for crops of rice, with at least 20% of this area subject to systematic crop failures due to a physiological disease of rice, which has been called "akiochi" (Takijima, 1963). This disease is reflected in the fact that there is a good first growth of rice, and by the end of summer the grain yield is suddenly reduced. It occurs on highly humic, peaty, poorly drained soils and marshy sandy soils. Studying this problem for a number of years, Y. Takijima found that damage to the root system is due to poisoning by reduced and ferrous products appearing under poor aeration and a high content of mobile organic matter in the soil. Around the roots, there forms a "cover" of ferrous iron salts; the ratio of acid-soluble ferrous iron to all the iron in the middle of summer is increased to 70%, while the roots have a very toxic ratio of about 8%. Hydrogen sulphide content in these soils, has double peaks (in July and August) and also exceeds the threshold of physiological stability. Among the 10 identified organic acids in soils, acetic and butyric acids occurred in the largest quantities, the content of which also peaked at the end of summer. Improving soil aeration eliminated its toxicity.

Along with this, there is a soil toxicity arising in the ordinary way and in the gaseous regime there is the accumulation of harmful waste products of plants, microorganisms and animals. This kind of toxicity underlies the phenomenon of soil sickness and is directly related to the subject matter of this book. The problem of allelopathy cannot be reduced to soil sickness. It is much wider, primarily because cases of soil sickness most often are specific. For some species, the soil is fertile, for others it is not. In addition, among the causes of the phenomena encountered as soil

---

[10] Grodzinskii does not elaborate about manganese, which under reducing conditions is not oxidised, but accumulates as Mn^{2+}, which in large amounts can be harmful especially to shoot growth.

sickness, there are those not due to allelopathy; unilateral removal of nutrients, reproduction of pests and pathogenic organisms, etc. Finally, the accumulation of cholines in the environment is not only regarded as an obstacle to the growth of plants and reduces their productivity, but as conditions favour vitality, it enhances growth and raises productivity. In addition, allelopathy is dominated by the impact of growing plants at the same time, in which the accumulation of cholines in the environment is temporary and generally may not be significant, since they are immediately consumed by other plants.

The accumulation of physiologically active substances in different soil layers is due to the presence of available organic matter (excretions) and microbial activity, which, on the one hand, involves the active excretion of plants, and on the other – the self-same active substances produced. As stated by microbiologists (Mishustin, 1961, 1963)[11], in all forms of tillage of the soil, the individual horizons of the arable layer differ significantly in their microflora and energy of the processes occurring.

The highest metabolism and activity usually characterises the upper soil horizon, due to better conditions for aerobic decomposition of various organic residues. The difference between the upper and lower horizons appears most clearly if there is no ploughing or subsurface ploughing by the method of T. S. Mal'tsev[12]. When ploughing with turned-over furrow slices, the residues of the last year fall to the bottom of the furrow. In this case, in the soil profile there is a second maximum of the microflora.

These data guide us in conducting direct measurements of choline accumulation in soil and in other structural parts of phytocenoses. Such determinations are made by relatively simple sampling in cenoses and the corresponding tests of volatile substances or aqueous extracts are accomplished by biological methods.

In the natural conditions of the cenosis, dead aerial parts typically comprise the surface of the litter, and rarely are found in the soil. The value of litter in the creation and improvement of water-physical properties of the soil is well known. The allelopathic value of litter is often underestimated. In experiments on sod-podzolic soil, we never received sustained negative effect of mulch on plant growth. So, mulches of dried and fresh residues of Tatarian bread and mugwort increased, in one of the experiments, the weight of oat plants by 425%. In a two-year experiment with brome grass (*Bromus inermis* Leyss.), all varieties of mulch (12 different plants) stimulated the growth of the brome grass. There was especially strong stimulation of growth with the mulch of Crambe (250%). On black soil and sand, mulch often has a detrimental effect on the growth of the brome grass and oats.

Consider the physiological activity of compounds formed by the decay of various plant materials.

N. Mokhov in our laboratory has carried out the following experiment. 10 g air dried material was moistened with 10 mg of water and composted at 20-22°C for 10

---

[11] These two references were omitted from the Reference List, and thus we have added the two most likely items there.

[12] It is worth noting that the techniques of Terence S. Mal'tsev (1895-1994), a Soviet agricultural pioneer, were important in Grodzinskii's postgraduate work.

days. Thereafter, the activity of volatile substances was tested on germination of radish (Table. 24).

Table. 24. Activity of miasmins in various plants

Plant species	Sample	Germination of radish at K-50%
Black henbane (*Hyosciamus niger* L.)	Above-ground parts	3.7
Peppergrass (*Lepidium ruderale* L.)	Above-ground parts	21.2
Black nightshade (*Solanum nigrum* L.)	Above-ground parts	23.2
Thorn-apple (*Datura stramonium* L.)	Leaves	34.6
Canadian fleabane (*Erigeron canadensis* L.)	Above-ground parts	35.3
Common wormwood (*Artemisia absinthium* L.)	Above-ground parts	47.0
Fat hen (*Chenopodium album* L.)	Above-ground parts	55.7
Mugwort (*Artemisia vulgaris* L.)	Leaves	65.5
Barnyard grass (*Echinochloa crus-galli* L.)	Above-ground parts	87.0
Wild mustard (*Sinapis arvensis* L.)	Above-ground parts	88.4
Winter wheat (*Triticum aestivum* L.)	Stubble	103.7

It can be seen that the volatile emissions from the rotting matter - miasmins - are very active in bioassays. Moistened strips of indicator paper were exposed to vapour of the miasmins, and indicated alkaline pH. Determinations with Nessler's reagent indicated that the pH may be caused by the presence of ammonia. However, there does not seem to be a direct relationship between the pH, the presence of ammonia and the strength of inhibition. Obviously, in the content of miasmins, other substances are also present.

This experiment was repeated with different conditions of composting, with different plant materials; the effects of miasmins were studied not only on seed germination of radish, but also on other plants, and on growth of stems and roots, respiration of seedlings, guttation, permeability of the protoplast and other indicators. Marasmins have been found to be extremely widespread; they are produced in almost all cases in the decomposition of any plant material. The dynamics of emission of miasmins during composting plant material in non-sterile conditions with the free reproduction of microorganisms has a peculiar character. Along with overall downward trends in activity and transition to stimulation, there occur at times sharp increases or decreases. In general, the curve has a wavy or sawtooth character. Apparently, this is due to shifts in the microbial cenoses that gradually decompose the material. The appearance of each new group of microorganisms is accompanied by destruction of toxins that has accumulated due to the previous group, and the gradual accumulation of new toxins, leading to the next change of microbial community.

There is a sharp difference between individual donors of miasmins. Thus, some of the plants completely inhibit the germination of seed, while others stimulate growth in higher ranges (up to 160% of control).

Studying miasmins is of great interest. According to our definition in terms of natural cenoses, miasmins are constantly held and given off by the soil, dead litter, forest or grass litter. They saturate the soil and the atmosphere, affect the germination of seeds and the vital activity of root systems (Samtsevich, Borisova, 1963).

Soluble phytotoxic substance released during the decomposition of plant material (saprolins), or produced as the physiologically active secretions of fungi and bacteria, also seem to play a huge phytocenotic role.

In our experiments, it has consistently been found that from the outset of decay, the amount of extracted phytotoxic soluble substances increases, and stays at that level for quite some time. For example, in one of the experiments, the composting of plant materials over 67 days led to a gradual increase in activity of saprolins to the end of experiment in samples of oats and wheat straw and leaves of *Datura stramonium*, and in some, to a small decrease in activity in two different samples of wild mustard. The composting of the residues of 10 species of plants for 110 days at 50% relative humidity and room temperature led to results that are given in Table 25.

Table 25 shows that in most cases the activity of saprolins decreased compared to the initial activity of cholines, unaltered plant substances, but still at the end of the experiment was quite appreciable. In a sample of the aerial parts of lambsquarters and wheat stubble, the activity increased significantly. Radishes and poppy in most experiments responded to saprolins similarly.

Table 25. Activity of aqueous extracts before composting (diffusates) and after 110 days of composting (saprolins) in relative units

Sample	Diffusates	Saprolins
Winter wheat stubble	11.7	20.7
Buckwheat stubble	71.0	13.0
Barnyard grass straw	101	30
Broad bean stubble	137	39
Common amaranth	176	48
Oats straw and stubble	250	205
Peas, residues from combine harvesting	280	160
Wild mustard straw	358	96
Lambsquarters straw	495	1100
Field bindweed residues	720	34

Although we composted relatively small number of species (60), mainly grasses, the results of the determination of inhibitory activity is so overlapping that it is possible to assume that the formation of saprolins is like that of miasmins, in any dead plant material.

All these examples show that any organic material released to the conditions of the phytocenosis, is the source of a variety of physiologically active compounds, that are harmful and useful for higher plants.

From this, the conclusion follows in favour of our assumptions about the significant phytocenotic role of plant excretions. Many phytotoxic substances, that are found in plants or in their secretions by biological tests, have been proved to be active in relation to whole, normally functioning plants in water or sand cultures. Soil is an extremely powerful factor in the environment, weakening the effect of phytotoxic substances, but, on the other hand, the soil is an accumulator of the same substances, and prevents their loss through evaporation or filtration of seepage water. At sufficiently high concentrations of toxic substances, the depressive effect in the soil is detected, as well as in water or sand cultures or in bioassays.

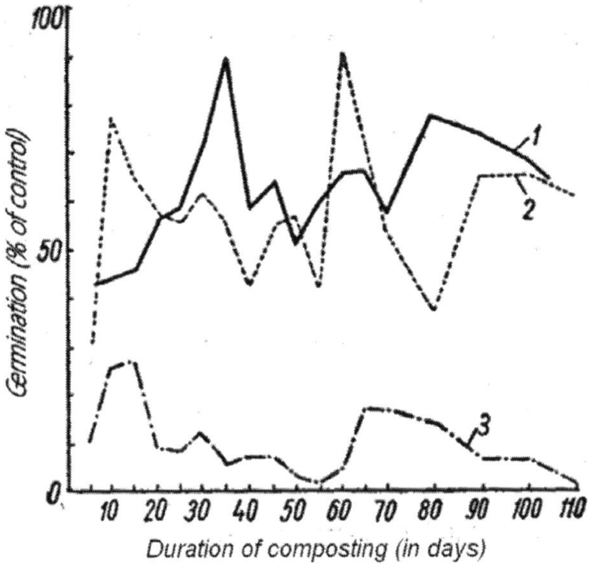

Figure 22. Change in activity of saprolins with time of composting of plant materials: 1 – *Echinochloa crus-galli* 2 – *Amaranthus retroflexus* 3 – *Chenopodium album*

Usually the plant organs that are most affected, have immediate contact with the active substances, while those that are more remote can simultaneously have enhanced growth.

The phenomenon of allelopathy should be viewed logically; at high concentrations or prolonged action suppression occurs; at low concentrations or short-term exposure one finds a stimulating effect of toxic substances *in vitro*.

The essence of allelopathy is thus reduced to the dynamics of active substances in the cenosis.

Chapter 4

THE ABSORPTION OF ORGANIC SUBSTANCES AND THEIR PHYSIOLOGO-BIOCHEMICAL EFFECTS ON PLANTS. THE PLANT AS AN ACCEPTOR OF ORGANIC SECRETIONS

In the realisation of chemical interaction, the response of other plants to substances secreted and contained in the environment is very important. Any activity of allelopathic compounds is dependent on the ability of plants to absorb these substances, accumulate or recycle them in their body, and thus ultimately, the effectiveness of allelopathy.

Therefore it is very important to consider the ability of a higher plant in the absorption of organic products, particularly the speed and dimensions of the process. Next, one needs to understand the issues about the mechanisms and methods of stimulation, and the toxic effect of active substances in retarding the various physiological processes in plants, their growth, productivity and quality of the crop.

Finally, we must consider the interaction of allelopathy and competition or commensalism, i.e. determine the relative importance of these types of mutual influence of plants and establish the relationship between them.

Release of organic substances in plants. Exchange of substances between plants growing nearby

The effect of any substances on the internal physiological processes is possible only when these substances are received into the body and there is assimilation into the protoplasm. A. Winter (1960) believes that allelopathic interactions begin with the moment of absorption of the physiologically active compounds by acceptor plants, even if it does not lead to significant changes in metabolism. The very fact that there is a physiological effect of a given substance demonstrates that it has already entered the plant.

Numerous data indicate that the plants easily absorb quite complex organic compounds. Curiously, the idea of organic plant nutrition arose before that of mineral nutrition, and formed the basis of the so-called humus theory.

Since the discovery of photosynthesis by Ingenhousz in 1779 and further development of research on autotrophic nutrition of plants in the works of N.-T. Saussure, J.-B. Boussingault, Timiryazev and many other scientists to this day provide a clear and indisputable idea that green chlorophyllous plants themselves create the organic matter that makes up their bodies. Furthermore, green plants are almost the only source of organic matter for all other living organisms, except chemosynthetic bacteria for which the total productivity, however, is very small.

Now we cannot doubt that plants are able to use water, carbon dioxide and mineral salts to create by means of assimilation in sunlight any of the complex organic compounds, which are known in nature.

With the development of these concepts in the first half of the last century, the ideas of A. Thaer and J. H. Hassenfratz on plant nutrition through the organic matter of soil were promoted. These ideas, called humus theory, came primarily from the fact that the yields of plants are generally bigger, when the soil humus is richer, and from the fact that manure and other organic fertilisers (mineral fertilisers were then almost unknown) increase productivity. On the basis of this theory, agriculture, crop rotation, etc., were developed.

Following the development of J. von Liebig's theory of mineral nutrition, and after studying the main features of microbiological processes in the soil, it was possible to explain all the phenomena of life and the productivity of plants, from the viewpoint of photosynthesis, inputs of salts and water, and mineralisation of organic matter in the soil with the help of microorganisms. Humus was assigned the role of the colloid-adsorbent that improves the hydro-physical and nutritional regime of the soil. It all seemed quite reasonable and suited many. In this regard, the humus theory lost recognition and was subjected to withering criticism, although some of its supporters struggled on until the end of the last century. In particular, J. von Liebig himself, although his work dealt a fatal blow to this theory, could not completely give it up, and believed that in the basis of humus theory there are correct observations. In the second half of the last century L. Grandeau made an attempt to combine the theory of mineral nutrition with humus theory. In the last years of his life J. von Liebig described this attempt as laudable, but it was unsuccessful. This eclectic mix of various propositions found no supporters and was soon forgotten.

Meanwhile, until recently, the role of organic nutrition involving plants is unduly small, although it is certainly important in the life of plants and can therefore be of great practical importance. We have a good idea about the uptake of cations and anions through the roots and foliar feeding, but still cannot give a satisfactory explanation for the beneficial effect of organic and organo-mineral fertilisers in comparison with pure minerals.

A review of the literature on plant nutrition with available organic compounds was published in 1911 by Hutchinson and Miller, and then there is some material collected by N.A. Krasil'nikov (1958).

The value of organic food in the life of the plant should be evaluated properly. Indeed, we can get a good harvest in a fully mineral medium, but in the wild plants always grow on a substrate with a certain amount of organic matter. The amount of so-called soluble humus in the soil, as discussed previously, may reach 1-2% by weight of soil, i.e. up to 100 tonnes per 1 hectare.

There is a legitimate question as to whether the plants cannot use these stocks as a source of organic food? How important is the organic food for plants in nature to produce the mass of yield, and how is it especially important in determining the quality of this mass? We do not know yet.

We are particularly interested in the problem of organic nutrition of plants due to the chemical interaction of plants, the fact that any substance secreted by a plant can act on itself or a nearby plant, essentially, only when it is absorbed from the environment. If this absorption is absent, then the effect of the substance will be weak or will not appear.

From experiments with isolated roots, it has been known for years that they can be cultured on a nutrient medium containing sources of organic matter, sugar, certain amino acids and vitamins. In experiments with cultures of isolated tissues, it was shown that they can use as a carbon source, not only the mono- and disaccharides, but also oligosaccharides and even starch (Gautheret, 1959). This indicates a relatively high extracellular carbohydrase activity. Numerous experiments have shown that during the growth of the explants on a medium containing starch, products of hydrolysis appear later. In the culture of viral tumors of sorrel (*Rumex acetosa* L.) the presence of extracellular α-amylase was immediately established. It hydrolyses not only starch, but also sucrose, maltose, raffinose (Constabel, 1963). High activity of invertase was found in the cell walls of explants from many plant species (Straus, 1962).

Bare roots are able to absorb from the soil amino acids and those vitamins, that they are unable to synthesise, in particular, nicotinic, pantothenic acid and the like. Metabolised by roots, amino acids are subjected to oxidative deamination with the release of ammonia, which is then used for the synthesis of new amino acids and amides (Ratner, 1963). The influx of assimilates from the aerial parts enhances this process, and possibly the influx of corresponding enzymes, such as glutamine synthetase.

Under sterile conditions, the ability of plants to absorb some form of organic phosphorus compounds has been demonstrated (Ratner, Kolosov, 1954).

In experiments with cultures of isolated carrot roots, it was found that they use organic sulphur compounds - cystine, cysteine, methionine. The sulphur-containing amino acid taurine was toxic for tissue cultures (Heller, 1955). The ability of adult plants to absorb methionine from nutrient solution was demonstrated by A.V. Manorik (1958) using a labeled compound.

These data indicate that plant roots can absorb conventional compounds such as amino acids and sugars. Clearly, these substances, after penetrating into the body of the plant, are no different from the same compounds that are formed during photosynthesis. They are used in various metabolic processes to build the body of plants.

As for the other more specific substances, it was also proved that there is their accumulation in plants after root uptake. So, entomologists have long established the penetration into plants of some organic insecticides - hexachlorane, DDT and others, to such a degree that these plants are toxic to pests (Bogdarina, 1961). It has been shown that the absorption of iron and other metals occurs in the form of complex organic systems - chelates, where during the process of absorption and translocation within the plant, the chelate molecule is not destroyed.

Entry into the plant was found for a number of antibiotics that are produced by microorganisms (Brian *et al.*, 1951; Blanchard, Diller, 1951; Winter, Willeke, 1952), such as penicillin, aureomycin, streptomycin. A. Savulescu *et al.* (1963) have established the root uptake by tomato of dihydrostreptomycin, biomycin, arenarin. N.A. Krasil'nikov (1962) found in sterile cultures that the juice bled from various plants contained almost all the antibiotics that had been added to the nutrient solution.

Antibiotics in most plants do not change and can be detected in the guttational liquid.

G. Franz (1962), and before that, A. Winter found that wheat plants and beans absorb from the nutrient solution a variety of alkaloids. Most of them are first accumulated in the roots, and in the leaves remain unchanged. The authors suggest that the flow of alkaloids occurs purely passively, together with a flow of water, but the adsorption of alkaloids, for example berberine, also occurs on the root surface.

Alkaloids are also absorbed from the soil. In one experiment we grew wheat in the sand, to which was introduced aboveground mass of Jimson weed (*Datura stramonium* L.) and black henbane (*Hyoscyamus niger* L.), which contain hyoscyamine. This alkaloid was then chemically identified by us in the roots of wheat, but we did not find it in the aerial parts. G. Franz (1962) suggests that quinine is stored in the tissues of wheat for a few months. Plants that are constantly receiving alkaloids from the environment, destroy them faster than those plants that got them for the first time, indicating that the formation of the enzyme complex is adaptive.

Besides alkaloids, plants can assimilate the products of their bacterial destruction.

It has also been established that there is the uptake of a number of phenolic compounds in plants (Winter, Schönbeck, 1959 a, b; Winter *et al*, 1959). This fact is particularly interesting, because phenolic compounds are formed during the decomposition of straw, dead leaves and other debris, and play a crucial role in allelopathy. We found that most of the phenols entering into the plant are rapidly degraded or undergo glucosylation.

According to F. Schönbeck, A. Winter (1962), to detect phenolic substances taken up by plants, from washed residues lying on the soil mass, they should have a weight of the order of 1-2% by weight of soil in the vessels. Further, they found that seedlings of wheat and beans absorb from nutrient solution coumarin, umbelliferone (7-hydroxycoumarin), esculetin (6,7-dioxycoumarin), scopoletin (6-methoxy-7-oxycoumarin). Because these substances, like phenols, have an OH-group, they may be glycosylated in plants, and be more mobile, and accumulate.

A. A, Chasovennaya (1961) found that the volatiles substances of giant hogweed (*Heracleum mantegazzianim* Somm. *et* Lev.) penetrate into the leaves of the test plants and cause damage to the protoplasm, and to chloroplasts. In this regard, at night, when the stomata are closed, the toxic effect of volatile production does not occur.

Many authors believe that humic acids enter the plants (Kononova, 1962). This is difficult to judge, since there are no methods of chemical identification of humic substances in plants. Pagel' has shown that with the penetration of various fractions of humates in plants, in which he determined the strength of the physiological effect, it was inversely proportional to the size of the molecules.

Numerous facts described in the literature relate to changes in the quality of plants growing in the neighborhood of other plants. N. Cézard (1960) describes the case where the wine produced from grapes grown near the hedge of blackberry, had in its bouquet the obvious smell of currant.

These facts indicate that the plants can be supplied with organic substances that occur in the soil. Absorbed compounds play a structural role, being a part of the substances from which the plant builds its body. In addition, they are growth regulators, which are able not only to stimulate, but also to inhibit the growth of plants, or even cause their death.

Experiments on the uptake of organic matter have been performed on a variety of plant species belonging to different families. Thus, higher plants are able, to some extent, to engage in heterotrophic nutrition, even in saprophytism. The lower green plants are known to be very inclined to food prepared with organic compounds. Thus, Euglenophyta lacking chlorophyll have many forms, which use organic substances dissolved in water as a carbon source. Chlorella and other obligate green algae develop very well in water, rich in soluble organic matter. Under the influence of organic substances: sugars, amino acids and others, algae reduce their rate of photosynthesis, i.e. organic food replaces photosynthesis (Litvinenko, 1960).

Unfortunately, insufficient attention has been paid to the organic nutrition of plants. Despite the successes of hydroponics and growing crops on synthetic media, product quality is often not the same as when growing plants in natural soil. Certainly, it is possibly due to the use by the plant of metabolites from the soil microflora, excretions of neighboring plants, humus, and so forth.

The mechanism of absorption of complex organic substances by plants, in particular, has hardly been studied. Among the substances listed before are those that are in nature: neutral compounds (sugars, carbohydrates), acidic, basic (alkaloids), and amphoteric (amino acids); among them one type dissociates into ions, another is hydrolysed by enzymes, and the third enters unaltered. If we accept the cellular structure of the protoplast in accordance with Ruhland's theory of ultrafiltration, then it is unclear how there is the relatively easy flow of huge organic molecules. However, some German researchers supporting this theory, believe that neutral substances have no chemical or electrical affinity to the surfaces of ultrapores in the protoplast and are drawn passively with water flow. In this case, the speed of their inflow is higher, with the smaller the size of the molecules, as evidenced by direct determination. However, this relation is easily explained by differences in the diffusion rates of different molecules in the surrounding solution, and dead zones of root cells.

The theories of "carriers", "free and inner space" and "anionic respiration" are currently common abroad, regarding the supply only of mineral ions, and do not consider the possibility of the uptake of organic substances in plants. This, incidentally, is a significant drawback in all of these theories and hypotheses.

Ideas about the mechanisms of absorption of substances, as put forward at onc time by D. A. Sabinin (1955) and developed by I. I. Kolosov (1962), may be used to explain the mechanism of entry of organic substances. According to this theory, the first stage in the uptake of substances is by adsorption on the surface of the root, that may apply equally not only to the mineral, but also to organic substances. The differences may be due only to the diffusion rate and the nature of adsorption exchange with substances already absorbed on the surface or secreted by the root. The next stage in absorption is in the metabolic binding of substances absorbed by the cytoplasm of root hairs. Molecules of sugars, amino acids, lipids, proteins and many other substances, that occur in the cytoplasm or some parts as with endogenous substances, are simply "embedded" in the protoplasm, but not filtered through some membrane. Apparently, the process of uptake of substances occurs using many very different mechanisms.

The mechanism of input of organic matter becomes especially clear when we remember that the root is a heterotrophic organ and its cells derive organic matter from other parts of the plant. In principle, each plant cell is able to take in organic molecules, otherwise its existence would be impossible.

It is not necessary that the incoming substance is a permanent component of the protoplasm of the plant. In the presence of extracellular enzymes, and the so-called "boundary-digestion", substances are recycled and are involved in the exchange of substances.

However, the plant absorbs, as we have seen, substances that are unnecessary and even harmful to it. The reason for this, as it seems, lies in the fact that such substances in some part of the molecule resemble the structural material of protoplasm. Therefore, they may come into temporary connection with certain proteins in the protoplasm, and thus enter the plant. In the case of toxic substances, they will "find their way" dramatically increasing the permeability of the protoplast, damaging cellular elements, as they penetrate deeper into the plant.

Thus, the question of nutrition of plants with prepared organic compounds can be considered to the extent of drawing the attention of physiologists to broaden and deepen work in this area.

In order to determine the allelopathic and ecological role of organic nutrition of higher plants, it is necessary to quantify the consumption of these substances.

Virtanen (1955)[1] grew plants under sterile conditions on extracts from peat and found that 1.8-5.3% of the total carbon content of the test plants was due to organic matter in peat extract.

A very vivid picture of the uptake of excreted plant substances into neighboring plants is given with experiments using labeled carbon. Extremely interesting work was done by V. P. Ivanov and co-workers (Ivanov, 1962, 1963 b, Ivanov *et al.*, 1963). They fed labelled carbon dioxide to leaves of either corn or broad beans, sown together, in a pot experiment, and determined the speed, methods, and the size of the movement of carbon through the roots and aerial parts of the plants.

---

[1] No reference is provided for this citation, and it would seem that this refers to work by the Finnish worker Olli Virtanen.

Obvious amounts of carbon in the neighboring plants were detected within 12-16 hours; the movement carried out by the roots, and through them to the aerial organs. Curiously, the tissue of the nodules on the roots of beans had 13 times the radioactivity through the excretions of corn, than tissue from the roots of beans themselves. This supports the fact that a neighbouring plant can supply carbon as food to the nodular symbionts of broad bean plants.

In further experiments, V.P. Ivanov also showed that the transfer of labelled carbon dioxide could occur also in the aerial parts, under static atmospheric conditions. In the open air, such exchange was virtually nonexistent.

With the participation of M. Savchenko, we repeated the experiments of V.P. Ivanov with various plants grown in aqueous culture and in the field, and we fully confirmed his findings.

In a field experiment with mixed planting of lupine and oats, it was observed that plants of one species absorb less or not any labeled secretions, but representatives of the other species, on the contrary, absorb them more strongly. This pattern is confirmed in some experiments with C^{14} with other types of plants in the field.

In July and August 1962, plants growing together in various combinations were selected. The aboveground part of a plant (donor) was placed in a chamber for photosynthesis for 2 hours, and was administered labelled carbon dioxide. Donor plants and acceptors were excavated at 24, 48 and 72 hours, dried and used to make autoradiographs. The study of the autoradiographs revealed the following:

1. The root exudates of donors included organic compounds with C^{14} assimilated during the process of photosynthesis, and possibly carbon dioxide, which was released during respiration of roots.
2. These substances in soil conditions were assimilated by the root systems of neighboring plant-acceptors. Transmission occurs quite rapidly and within a day or two, there were significant amounts of C^{14} in plant-acceptors.
3. Transfer from crop plants: pearl millet, lupine and oats, to weeds: barnyardgrass (*Echinochloa crus-galli* (L.) Roem. *et* Schult.), couch grass (*Agropyrum repens* (L.) P. B.) and yellow foxtail (*Setaria glauca* (L.) P.B.) was very good and; on the contrary, from the weeds of crops it is very weak (e.g, foxtail to oats, and from foxtail to lupin), or did not occur (from wheatgrass to lupin).
4. The exchange between different types of weeds (odourless chamomile, barnyard grass, couch grass, shepherd's purse) was vigorous. The intraspecific exchange of weeds in some cases was strong (chamomile among themselves, and barnyard grass with themselves), in other cases weak (foxtail with foxtail, and couch grass with couch grass).
5. Among the crop plants, the exchange between plants occurred very well in vetch with oats. In corn and beans, and among the plants of corn in this phase of development, there was no exchange.
6. In all cases, the closer the donor and acceptor were positioned, then the better was the transfer of root exudates that occurred.

In experiments with water cultures, groups of two-week old plants (wheat, rye, oats and barley), which have been grown in containers on Helrigel nutrient solution, were taken. Another group of plants was five weeks (old).

Donors were fed $C^{14}O_2$ in a chamber for photosynthesis, then remained for 24-hour in nutrient solution in a separate room. Then they were positioned in their place among other plants. After 24 hours, a portion of the shoots was collected, and after 48 hours also were the remaining stems and the entire root system of the acceptor plants. Data on the radioactivity from a series of experiments are given in Table 26.

Analysis of the results obtained in experiments with water cultures, makes it possible to draw the following conclusions:

1. Part of the photosynthetic products containing labelled carbon, was found in the contents of the root exudates, in the nutrient solution, and was taken up by other plants.
2. After 24 hours C^{14} products in large quantities were detected in stems of younger acceptor plants; but five weeks after planting, no radioactivity was detected. The uptake of labelled products into old acceptor plants was slower than in young plants.
3. The radioactivity of the aerial parts of acceptor plants was 2-7% of the activity of the aerial parts of donors but the roots of acceptors have radioactivity that sometimes attains 30% of the activity of the roots of the donors. C^{14} total transmission was 2-8 % of the amount of carbon assimilated by donors during photosynthesis.

Table 26. Transfer of the labelled products of photosynthesis through roots in aqueous culture (radioactivity of samples in counts/min per mg)

Donor			Acceptor			
Plant	Sample	Radio-activity	Plant	Sample	Radio-activity	% radioactivity of donor
Oats	Aerial parts	452	Oats	Aerial parts	12	2.7
	Roots	209		Roots	58	27.6
Wheat	Aerial parts	493	Wheat	Aerial parts	1	0.2
	Roots	169		Roots	22	13.0
Rye	Aerial parts	446	Rye	Aerial parts	3	0.5
	Roots	209		Roots	19	9.0
Wheat	Aerial parts	430	Rye	Aerial parts	3	0.7
	Roots	180		Roots	30	11.5
Rye	Aerial parts	534	Wheat	Aerial parts	29	18.6
	Roots	220		Roots	31	13.9

In aqueous cultures, the transfer of labeled compound between different species and between individuals of one species occurred at about the same rate, when comparing plants of the same age. This suggests that the above-noted lack of transfers of fluids within a species may be due not so much to any special relative position of the root systems of plants of the same or different species. In principle, plants, unlike animals, can resorb their own secretions.

The relatively rapid movement of C^{14} products from donors to acceptors allows us to assume that organic substances that are part of the root exudates, are readily soluble, and contain mobile substances such as carbohydrates, organic acids, amino acids, and carbon dioxide, for which the possibility of absorption in higher plants is reported A.L Kursanov *et al.* (1962), A.M. Kuzin and V.I. Merenova (1953), V.L. Voziesenskii (1958) and others.

The results obtained indicate widespread exchange of organic substances of root excretions between plants, at a high rate and of quantitative importance to this process.

Earlier, we presented data on the movement of gibberellin from one plant to another. It is appropriate to recall the experiments of W.H. Preston jr. *et al.* (1954) on the movement of α-methoxyphenylacetic acid (MOPA) between plants in water culture. Appreciable amounts of MOPA are found in acceptor plants only 9 hours after application in lanolin paste to donor plants (bean sprouts), and which was 2-3% of the amount deposited on the donor. MOPA is not altered chemically, which excludes the formation of salts.

It was found that there was movement of MOPA from: beans to sunflower, pumpkin, buckwheat, and cotton; from pumpkin to beans, sunflower, buckwheat, and cotton; from sunflower to beans; from buckwheat to beans; from cotton to beans. There was no evidence implying movement of MOPA from corn to any plant. In planting new plants after those that had been treated with the stimulant MOPA, there was no transfer of it via the soil. Obviously exchange occurs on a thin film of water close to or near the location of the roots. About thirty different synthetic regulators have been tested, but none of them moved in such quantities as to cause a significant reaction in the growth of neighboring plants.

A later study by the same authors (Mitchell *et al.*, 1961) has shown that the excretion of MOPA by roots can reach 20% of the applied amount, whereas under the same conditions, only 3% of 2,4-dichlorophenoxyacetic acid was excreted.

The question of transferring the mineral nutrient substances from plant to plant has been better studied. Interesting work using radioactive isotopes was carried out by the American researchers John E. Kuntz and A.J. Riker (1956). They found that the mineral nutrient substances and spores of phytopathogenic fungi could be easily moved from one specimen of oak to another oak. According to them, in an oak forest, almost all the trees have roots that join together, and thus, among them there is a close exchange of water, nutrients and pathogens.

It can be considered reliably established that nutrients and water move freely through the fused plants of the same or different species (Rubtsov, 1950, Yunovidov, 1951, Beskaravainii, 1955). Fusion of the root systems of trees has been observed very often and is not a rare phenomenon.

However, allelopathy is more interesting as the exchange of substances between plants where the roots do not grow together. A massive study of this issue began in the 1950s, after the tracer method became widely available. Great work was undertaken by I.N. Rakhteenko (1963) in pure and mixed cultures at the

Kolodischanskoie Forest of Minsk Forestry, and plantations in the Botanical Garden of the Belorussian SSR. Studied tree species were aged from 1 to 20 years. They were growing on sandy, sandy loam and loamy soils: common oak (*Quercus robur* L.), small-leaved lime (*Tilia parvifolia*)[2], Norway maple (*Acer platanoides* L.), Norway spruce (*Picea excelsa* Link), larch (*Larix sibirica* Led.), Pennsylvania ash (*Fraxinus pennsylvanica* Marsch.), white birch (*Betula verrucosa* Ehrh.), willowleaf spirea (*Spiraea salicifolia* L.) and some species of herbaceous plants. Radioactive phosphorus was introduced by wetting the leaves, through holes made in the trunk, or via the isolated ends of roots. After two or three days, samples were taken from the neighboring plants located 0.25-6 m distant. Thus 156 trees were treated, and the movement of P^{32} was studied in 474 plants. In such a way, there was collection of fully reliable material; I. N. Rahteenko notes that under the same experimental conditions, the movement of phosphorus from the limes among oak is more intense than that from oak in lime, or from lime in lime. From Norway maple in oak, the labelled phosphorus also moves faster than from oak in maple, or in one instance from maple to another maple. In pot experiments, in some cases the same pattern was established: phosphorus movement from one tree species to another (e.g. from a lime in oak, from spruce in larch) occurred more intensively than the exchange between plants of one species (from lime to lime, from oak to oak).

It was also established that the transfer of labelled phosphorus from one plant to another occurs when there is contact of the root systems; in the absence of direct contact of woody plants, exchange was absent.

From the studies of I.N. Rakhteenko, it follows that the marked differences in the transfer of nutrients within a species, and between different species, may be due to different rhythms of the excretory and absorptive activity of plants. The point is that plants during the growing season may have different ratios of these two processes; in different species the processes of absorption and excretion may not coincide; in the same species they naturally occur synchronously, simultaneously. Therefore when there are combinations of plants, in which one actively secretes, and the other actively absorbs substances, the transfer will be very effective, and in the pair of plants, one will be much weaker.

The transfer of labelled phosphorus is mentioned even in the direction from tree species toward herbaceous plants (couch grass, Kentucky bluegrass), in the case where their root systems are in contact. However, as the author notes, the transfer can take place quickly and through the soil solution without direct contact of the roots when sufficiently close in their distribution.

In the work of other researchers (Shain *et al.*, 1957; Shain, Kashmanova, 1953; Shain, Nikolaev, 1962; Rachkov, 1962, 1963; Nikolaev, 1963) it has been shown that herbs and plants with adjacent growth, very quickly exchange various nutrients. The experiments were conducted on a number of plants planted with successively adjoining roots, so that they formed a kind of biological chain (see Appendix).

S. Shain established the very rapid movement of P^{32} from plant to plant. Thus, in the experiment with corn, a month later he discovered radioactivity in the seventh

---

[2] Now known as *T. cordata*.

or eighth row. Unfortunately, the author does not give figures on the amount of transmitted phosphorus. We also planted in a box with sand, corn plants separated by two layers of filter paper. Unlike S. Shain with feeding applied through the leaves, we gave P^{32} to plants through the germinal roots, placed in small glass vials. Eight plants for the first row received together 100 curies P^{32} in the form of K_2HPO_4. The experiment ended after 30 days. Data concerning radioactivity (% of the activity of the donors) are given in the Table 27.

Table 27. Transfer of labeled phosphorus and calcium in biological chains of corn

No. of row	1	2	3	4	5	6	7	8	9	10
Labelled phosphorus										
Content of P^{32} (as % of row 1)	100.0	0.504	0.238	0.128	0.081	0.029	0.016	0	0	0
Labelled calcium										
Content of Ca^{45} (as % of row 1)	100.0	12.63	1.16	0	0	0	0	0	0	0

As can be seen from Table 27, the radioactive phosphorus in corn, during a month, can move as far as, and including the 7th row, and even where the roots have contact with each other only through two layers of filter paper. Plants of all acceptor rows together contain about 1% of the phosphorus absorbed from plants of the 1st row. Obviously, the best contact of roots can achieve a larger extent of exchange. It will be appreciated that the phosphorus compound interchange occurs simultaneously in two directions, longitudinally (between the different rows) and cross-wise (between adjacent plants in a row), so that the exchange attains even a conditionally reduced lower percentage of the total content of the studied substance in the plants.

Results from a similar experiment with Ca^{45} are displayed in Table 27.

Unlike phosphorus, calcium moved only to two adjacent rows, but its amount was significantly greater. When moving from row to a row, the labeled calcium content decreased each time by about 10 times.

Once again, we repeated the experiments on the transfer of labeled phosphorus through the chain of corn in order to establish the role of guttation in this process. Given that there is close proximity of corn plants such that in the aboveground portions, the ends of the leaves enter into the neighboring rows and guttational water flows from them, it can be a carrier of P^{32}. In the experiments of S. Shain and other researchers, the aerial parts were not isolated.

The experiment was conducted according to the procedure, which included other options, in which only the roots were separated, and it was only via aerial organs that a small amount of phosphorus - about 0.03-0.16% of the total donor content of P^{32} was transferred. The guttational fluid was radioactive. This fact must be taken into account in such experiments, that transmission through the roots is up to 6.1% of the amount of total uptake. As in the experiment described above, the movement of phosphorus was of complex nature with a sharp drop in the beginning, and then a more or less shallow declining curve.

This is a very interesting feature of the migration of phosphorus that has been particularly well shown by L.D. Rachkov (1962, 1963). Thus, he cites the graph that shows that the movement of phosphorus became established in the model phytocenosis very quickly - within a few days, after which it is as if there is a defined level of active phosphorus in all acceptor tubes. However, the transferred mass of phosphorus forms a small peak, which reaches the tube, changes and then goes backward.

There is an idea, that in the presence of root secretions, there are some special mobile forms of phosphate, which occur as a few percent of the total phosphorus in plants.

L.D. Rachkov showed a similar movement in the components of a model cenoses with radioactive carbon, sulphur, cobalt-60 and strontium-90. In the movement of each element there are differences, but in general there are common features: high rate of migration and considerable quantitative extent.

All the above-mentioned facts are confirmed in the works of V. Nikolaev (1963), N.V. Timofeev-Resovskii (1962) and others.

The study of migration of organic and mineral substances is of great theoretical and practical interest, S. Shain pointed out that plants in a phytocenosis feed from the soil as a whole. There is some definite merit in his statement.

Physiological and biochemical action of allelopathic active substances

The physiological and biochemical mechanisms of action of various active substances have not been studied enough. Despite the relative protractedness of this issue, it seems so complex that there is not an opportunity to build a single scheme of the intricate biochemical actions of physiologically active compounds administered from outside into the plant. However, the natures of the effects of a variety of substances have a lot in common, at least from the outward manifestation of the response of the plant. This is due to the fact that virtually all substances that have a physiological effect in relatively low concentrations (oligodynamic substances) have a similar effect on the various manifestations of the vital activity of living protoplasm. With a high content of substances, there is observed a strong inhibition or death of living entities; with a gradual decrease in the concentration, inhibition decreases in an approximately linear logarithmic relationship to a certain point, when the effects of the substance are masked and growth does not differ from that of the control; further lowering of the concentration of a substance causes stimulation of vital processes, which then can be reduced to the normal with solutions at infinite dilution.

Aggregate shape of the curve of action of the active substance is shown in Figure 23.

These curves are obtained by determining the action of changing concentrations on seed germination, root growth, shoot photosynthesis and respiration, absorption of substances and other more common reactions of plants. Similar curves can be obtained by increasing the exposure time of the experimental object under study in an aqueous or gaseous mixture at the same concentration. Mutually similar effects are produced with the most diverse excretions of plants, extracts and vapours of dead plant matter and from the soil. Practically almost every inhibitor described by us, with decreasing concentration or exposure time can have a stimulating effect, or any stimulant, by increasing its concentration, can be converted to an inhibitor.

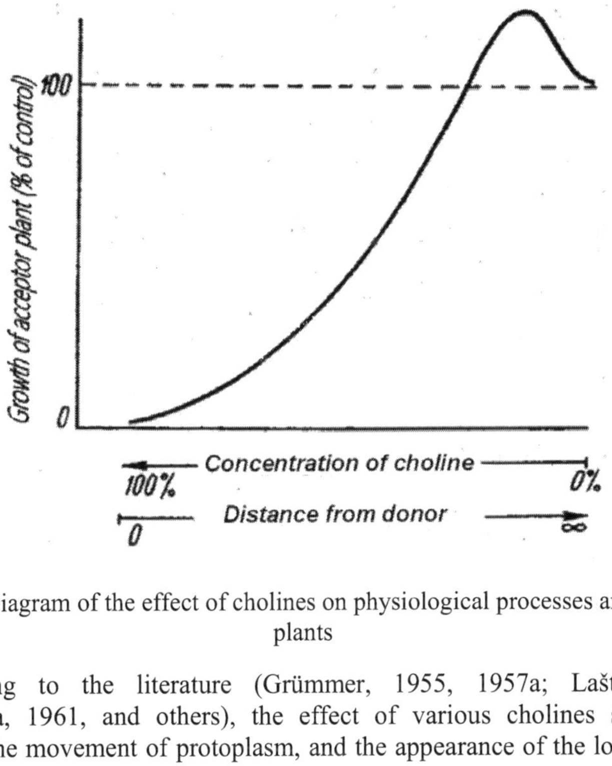

Figure 23. Diagram of the effect of cholines on physiological processes and growth of plants

According to the literature (Grümmer, 1955, 1957a; Laštuvka, 1961; Chasovennaya, 1961, and others), the effect of various cholines starts with a cessation of the movement of protoplasm, and the appearance of the loss of turgour, wilting plants, destruction of chloroplasts, accompanied by the appearance of chlorosis and anthocyanin coloring. Plant growth is slowed down and stops, and the plants die. These symptoms are accompanied by: severe disorders of carbohydrate and protein metabolism, and enzyme activity; and decrease in the uptake of nutrient substances, and photosynthesis.

All this suggests that the impact of allelopathic active substances on living protoplasm of plants is nonspecific.

According to D.N. Nasonov and V.Ya. Alexandrov (1940), the nonspecific response is in changing the degree of the fluidity of the protoplasm, in increasing its permeability, and finally, in the denaturation and coagulation of proteins. Since the external response of plants to allelopathic active substance is very similar to the nonspecific reaction, we decided to investigate the effect of plant excretions on the permeability of the protoplast and the capacity of proteins in hydration.

As a result of a large number of determinations, it was found that the volatile and water-soluble substances do not affect the swelling capacity of seeds of pea, corn, lupine, wheat and other plants, and therefore the delay or acceleration of germination cannot be explained by a change of the hydration properties of proteins.

However, the active substances fundamentally alter the permeability of the protoplast. The relevant data are given in Figures 24 to 27.

These data show that the response of plants to miasmins and other active substances is a typical non-specific reaction in the sense of D.N. Nasonov and V.Ya. Alexandrov.

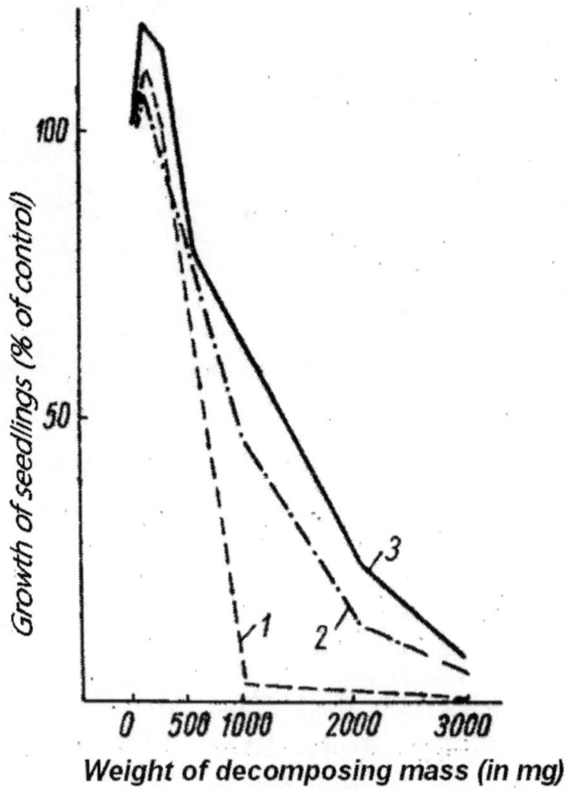

Figure 24. Influence of different exposure times to vapours of miasmins at differing concentrations on growth of cress seedlings.
1 - 24 hours; 2 - 48 hours; 3 - 72 hours

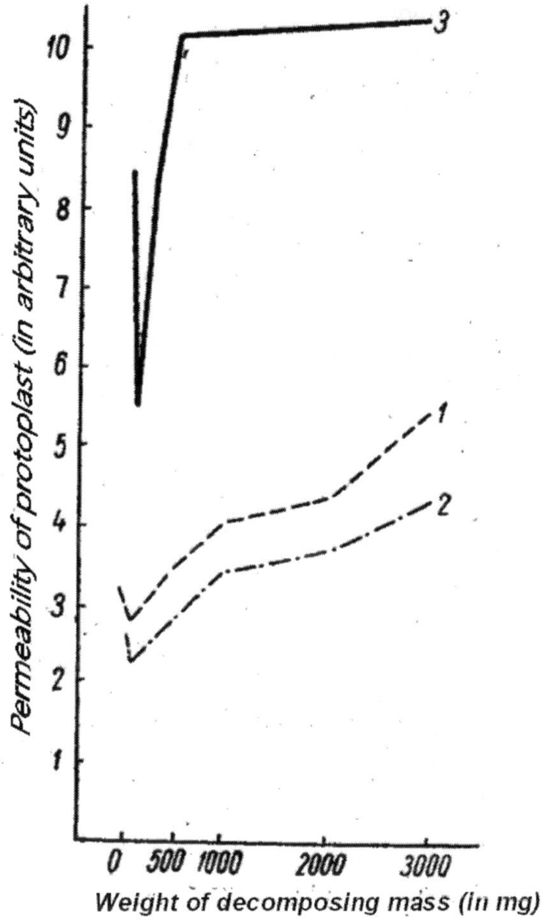

Figure 25. Permeability of protoplast in cress seedlings influenced by miasmins. 1 - 24 hours; 2 - 48 hours; 3 - 72 hours.

There is a relationship between changes in the permeability of cell membranes and the intensity of respiration and growth. Here one can agree with the indications of Sacher (1962), who believes that the pathological changes of permeability is likely to disrupt the concentration of salts in the protoplasm, change the pH, shift relations between the respiratory substrate and the enzymes, and cause the loss of a large number of metabolites to the environment, resulting, in the final account, for the progressive reduction of respiration.

Simultaneously with these are also changes in the absorptive capacity of the protoplasm, and, namely, the lesser the permeability of protoplasm, the higher appears to be its absorption capacity.

There is essentially an inverse correlation between changes in the permeability of the protoplast and retention, and the absorbent capacity. This state is contrary to popular opinion in the scientific literature that the increase in permeability of the protoplast is a positive factor contributing to a better flow of substances that enhance the hydrolytic processes, respiration, and so on. Obviously, an important role is played by the degree of change of permeability, not whether, on the whole, it increases or decreases. Moreover, this phenomenon is subject to fairly rapid change, so when using imperfect methods, errors may well occur. Change of permeability is a reversible process, if it has not gone too far.

It should be noted that one and the same concentration of cholines with various plant organs have differential effects. Aerial parts are more stable than the roots. As is known, auxins operate in the same manner. We believe that the reason for this is that the roots, the organs specifically adapted to absorb substances quickly, become saturated by cholines or auxins, and more affected by them than the aerial organs, which are better protected and absorb the substances more slowly.

Thus, for allelopathic active compounds, they are characterised by nonspecific effects. In the excretions of a wide variety of plants, there are very chemically dissimilar substances, which cause apparently the same reaction in plants. Geobotanists, occupied with issues of allelopathy, tend often to overestimate the specificity of action of some plant excretions in relation to others (Gortinskii, 1963). Meanwhile, what is more striking in these matters is the non-specificity of action rather than a definite confinement of excretions to suppress one plant and to stimulate another. The positive or negative impact of substances is more dependent on its concentration than on unknown specific "anti-forest", "anti-steppe" etc. properties. Using bioassays to simultaneously analyse a variety of processes, we always found similar results. For illustration, we present data on the seed germination of radish, spruce and oak in extracts from leaf litter of seven tree species; these data were obtained in our laboratory by F.E. Goloyadov (Figure 28).

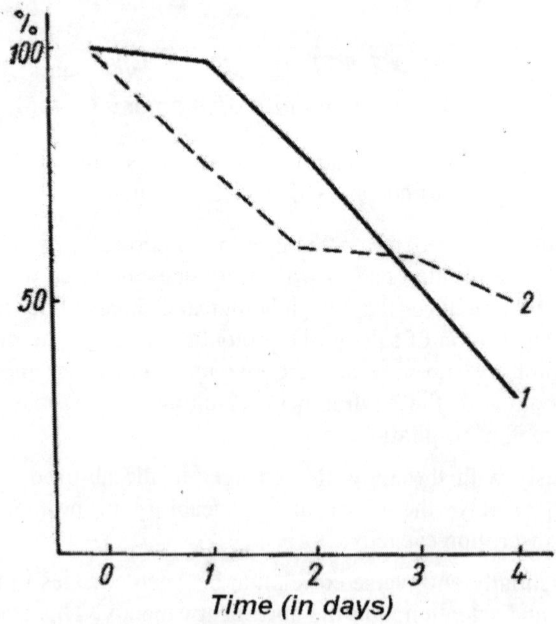

Figure 26. Effect of treatment on growth of (1) and respiration (2) seedlings of cress

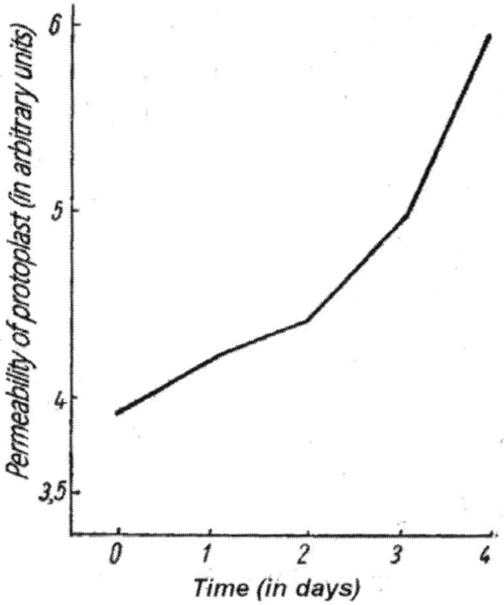

Figure 27. Changes in permeability of the protoplast by prolonged exposure to miasmins (same experiment as in Figure 26)

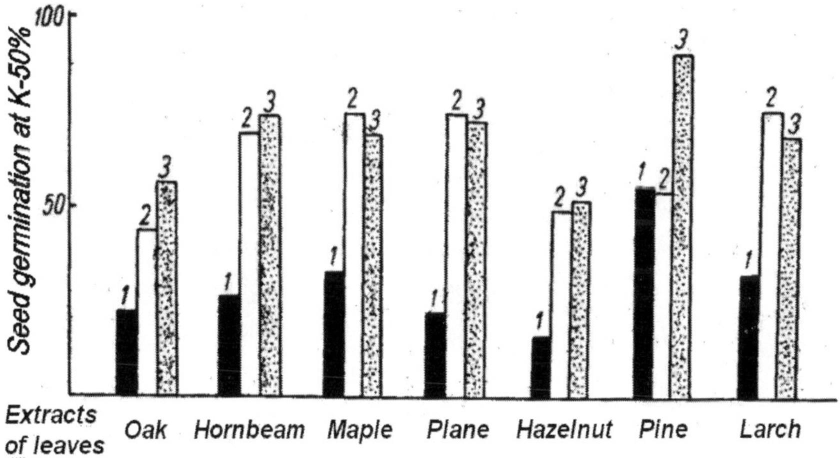

Figure 28. Effect of extracts from fallen leaves and needles of different trees on seed germination of radish (1) oak (2) and spruce (3).

The sensitivity of acorns, and seeds of common spruce to six extracts were almost identical and only to the extract of the fallen needles of Crimean pine was spruce somewhat more resistant. Radish is significantly more sensitive than spruce and oak. In obtaining seedlings, the same extracts were diluted 4 times. Meanwhile germination curves of radish exactly resemble the curves for spruce.

The general selectivity of the effects of cholines may be occurring, but it cannot be overestimated in experiments with bioassay, where all biological objects react very similarly.

The primary biochemical processes, subjected to cholines are almost unknown. From what we know from the literature and our own research, the physiological and biochemical effects of cholines can be classified roughly as follows :

I. Changes to physico-chemical properties of protoplast

(a) Enveloping membranes, clogging of pores, apertures, etc. (lipids, sugar, wax, etc);

(b) Changing the surface tension of the protoplasm (saponins, acids, alkaloids, tannins, ammonia, etc.);

(c) Modification of protoplast permeability (anions and cations);

(d) Plasmolysis (carbohydrates, acids, mineral salts);

(e) Dissolution of elements of structural protoplasm and nucleus (saponins, alkaloids);

(f) Denaturation of proteins (acids, bases, and other polyphenols.).

II. Changes in metabolism

(a) Competing with essential metabolites for places on carriers (structural analogues of metabolites);

(b) Loading of carrier molecules non-cleavable components (carbon monoxide, hydrogen cyanide);

(c) Competition with metabolites necessary for participating in the biosynthesis of (functional equivalents: antivitamin metabolites, antienzymes etc.);

(d) Induction of enzyme activity (substrate enzyme action);

(e) Depression enzymes (accumulation of the end products enzyme activity);

(f) Blocking of enzymes (heavy metals, organic acids, ureides, ammonia).,

(g) Blocking the metal (complexing - chelates, organic acid);

(h) Oxidation or reduction of the required metabolites (donors and acceptors of oxygen and hydrogen - polyphenols and polyquinones, humic acids);

(i) Change of oxidation - reduction conditions within cells (unsaturated fatty acids, terpenes);

(a) Blocking energy carriers.

III. Changing the physiology of the whole plant

(a) Changing the speed of water flow and nutrient substances through the roots; changes in the intensity of photosynthesis and other biosyntheses in plants;

(b) Strengthening or weakening of division and elongation of cells, plant growth and development;

(c) Anatomical and morphological changes;

(d) Decrease or increase resistance to adverse environmental conditions and pathogens.

All these modes of influence, the list of which could continue, have a place in the influence of gaseous or dissolved cholines. The study of the effect of toxic secretions of plants encounters some difficulties. One of them is the fact that under natural conditions, i.e. in the soil solution, in the gravitational and capillary water of the underground air, test substances almost never operate in a pure form.

There are always formed certain mixtures of various interacting substances secreted in the surroundings of other plants, animals, microorganisms. Among these substances there are considerations of antagonism or synergy, or that their action takes place independently (additive).

Despite the fact that the excretions of each plant comprise a very complex mixture of substances, there are found in various plants striking interactions of physiological properties of these secretions, which are often expressed as synergy, less so as antagonism, and even more rarely, as an independent additive effect.

Particularly strong synergism is observed between the secretions of plants that normally in nature form pairs, for example, between the oak and hornbeam, oak and hazel (Figure 29).

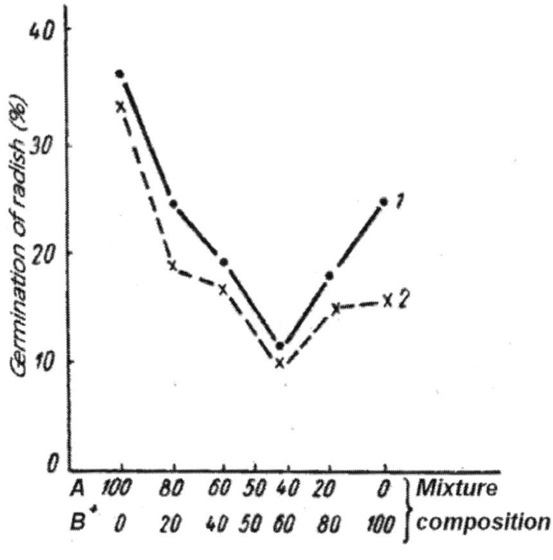

Figure 29. Synergy of extracts from fallen leaves of oak and hazel (1), and oak and hornbeam (2):
A – content in mixture of extract from oak; B – content in mixture of hazel or hornbeam

Plants, that occur together rarely in nature, have weaker syncrgisation of secretions.

When a third, fourth, etc. component is introduced into the mixture, the synergy is weakened and often replaced by antagonism. The effect of such complex extracts is weaker than ordinary. The same phenomenon was observed in natural conditions by our staff member V.K. Myakushko, who found that the activity of cholines in pure stands of trees is generally higher than in a mixed stand.

From the data on the interaction of secretions, it can be concluded that the phenomenon of synergy is not accidental. It could arise in evolution as an expedient tool.

The synergism of secretions shows that, despite the seemingly convergent nonspecific reaction of plants, the active substances affect completely different internal processes.

Consider the action of cholines on some physiological processes in plants.

The effects of various synthetic substances, phytoncides, saprolins, miasmins, root exudates and other cholines on the germination of seeds have already been repeatedly mentioned. It should again be denied, that cholines have any practical influence on the swelling of seeds, and only modify the energy of emergence or completely suppress germination at high enough concentrations.

Studies of the effect of root exudates on the mineral nutrition of plants were carried out in our laboratory by T.N. Filippovich (1962, 1963 a, b). The results of experiments with non-sterile and sterile cultures of various plants using the method of tracers showed that the uptake of phosphorus, sulphur, nitrate, potassium was highly dependent on the presence in the medium of root exudates from themselves, as well as other plants. At low concentrations, choline flow is enhanced by 1.5-2 times compared to the control; at high concentrations it is suppressed to almost zero. Substances that act on the root nutrition have the general nature of those substances that inhibit the germination of seeds.

In non-sterile conditions, the toxic effect of root exudates eventually weakens, which is associated with the destruction of microorganisms.

The absorption of substances by roots and their movement by aerial parts was delayed by toxic root exudates in different degrees. The first process proved to be more sensitive and reduced the intensity at weaker concentrations than the process of substances entering into the aerial parts, which are influenced by higher concentrations. This can be attributed primarily to the fact that the absorbing zone of the roots formed a kind of "cover" of organic substances that are part of the root exudates. Organic matter can be adsorbed on the root surface and prevent the penetration of ions, such as phosphate, into the root. As shown by I.A. Geller, D.A, Tabentskii (1957), the removal of this "cover" from the roots of corn using oxidants leads to an increase in intensity of the uptake of phosphorus.

The uptake of different substances, to differing degrees, depends on the action of root exudates. For example, concentrations clearly toxic for the uptake of phosphorus and nitrate were stimulatory for the flow of substances that to a great extent is determined by the rate of release in the roots and the rate of outflow from the aerial parts, but the ways of incorporating different ions in metabolism are diverse.

Root exudates of plants affect the flow of substances into neighboring plants also due to their chemical interaction with ions that are in the soil solution. Thus, the root exudates of corn, brome and quack grass have high nitrate-reducing ability. Acidic root exudates increase the mobility of potassium, calcium, phosphorus and other nutrients in the soil.

T.N. Filippovich obtained results that suggest the need to revise the concept of plant root competition for nutrients. All studied plants possessed to a greater or lesser degree toxic secretions and themselves to a greater or lesser degree altered the process of nutrition under the influence of root exudates. Therefore, competition always occurs against the background of chemical interaction that defines the whole process of the uptake of substances.

Detected dependencies can be very important for the process of control of plant nutrition. In particular, the good effect of organic and mineral fertilisers is usually explained by the fact that they increase soil microbial activity, resulting in accelerated mineralisation of organic matter and they improve plant nutrition. It is said that the microorganisms "feed" the plants with minerals, which, however, has not been confirmed by direct experiments. On the contrary, there is a fairly extensive literature that soil microorganisms often act as serious competitors of higher plants for nitrogen, as well as phosphorus, trace minerals and other substances.

Is it not better to assume that the beneficial effect of organo-mineral fertilising mixtures is due to the direct influence of physiologically active substances of the organic component of the mixture and excreting by activated microflora?

The effects of volatile substances on photosynthesis have been little studied, and in terms of allelopathy hardly studied at all. One can only cite the works by M.V. Kolesnichenko (1960 a, b. 1961 a., 1962, 1963) in which the definition of the mutual influence of woody plants, according to photosynthesis forms the basis of his method proposed for the co-cultivation of farmstead of forest species on farmsteads.

It is known that some volatile poisons such as hydrogen cyanide, carbon monoxide, hydrogen sulphide and others can have a dramatic effect on the rate of photosynthesis (Rabinovich, 1951). On the other hand there is a fairly extensive literature on the impact of industrial and transport volatile pollution on cropped and wild vegetation. Many of the facts can be attributed to the influence of volatile compounds on photosynthesis, which is a very sensitive process.

It is known that photosynthesis is the central process in the life of green plants, and therefore damage or stimulation due to allelopathic effect will have an impact on all other forms of vital functioning in the plant. It is also obvious that, among the other metabolic processes occurring in the leaves, photosynthesis is the most vulnerable, as it relates to the absorption of carbon dioxide from outside and it is in close contact with the external atmosphere.

In a number of experiments carried out in collaboration with I. N. Mokhov, we studied the effect of vapour from pulverised plant mass on the intensity of the absorption of labeled carbon dioxide by leaves. Photosynthesis was studied in constant lighting conditions, at a constant concentration of $C^{14}O_2$ exposure, and its intensity when treated with phytoncides of leaves was expressed as a percentage of the intensity of the absorption of carbon dioxide corresponding to the control.

Just as with donors, photosynthesis has been studied with more than 100 species of plants, and the acceptors were whole leaves and cut-outs from the leaves of beans, tomatoes, potatoes, corn and other plants. Phytoncides within wide limits changed the potential rate of photosynthesis – there were marked swings from 0 to 300% of the control. On average, the most resistant were cut-outs from the leaves of the potato, the most sensitive were from tomato leaves.

If we trace the process of action of phytoncides of various plants in dynamics, there are found clear, similar to each other, nonspecific changes in the rate of photosynthesis. In many cases, in the early treatment with phytoncides (within 15-30 minutes), there is stimulation of the absorption of $C^{14}O_2$, which is then followed by progressive inhibition (Figure 30).

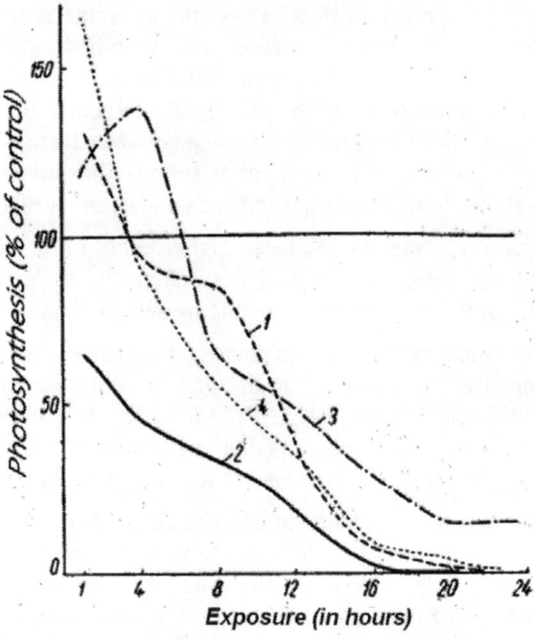

Figure 30. Influence of phytoncides of Russian almond on photosynthesis in test leaves with deffering exposure to phytoncidal vapours:
1 – potato, 2 – tomato, 3 – bean, 4 – corn

Based on these data, we believe that volatile substances affect photosynthesis primarily as strong irritants causing a typical parabiotic response. And then it begins to manifest itself as a more or less specific action leading to the progressive loss of photosynthetic capacity. There is the process of poisoning the photosynthetic apparatus. Possibly, after brief exposure to low concentrations, the process may be reversible.

By this means, the toxic effect of phytoncides of many plants on photosynthesis in the laboratory was very strong, and often led to death. In a natural setting such circumstances are rare. However, in nature the factor of the effect of continuously low concentrations comes into force. Our data imply that with increasing exposure, even to low concentrations of phytoncides, they gradually begin to suppress photosynthesis. Thus, the effect of volatile substances on photosynthesis in nature plays a very important allelopathic role, but the issue should be studied in more depth.

The influence of cholines on the water regime of plants has hardly been studied. Agropyrene excreted by couch grass, reduced the flow of water in wheat (Laštuvka, 1961). In the experiments of V.M. Gaidamak, working in our laboratory, he treated the aerial parts of wheat seedlings with miasmins; there was a 1.5-2 fold increase in the intensity of guttation. This is connected with an increase in the permeability of the protoplasts of the leaves, as the roots have been isolated from the influence of miasmins. The literature indicates that cholines increase water loss by plants. Changing the function often entails a change in structure. In particular, under the influence of cholines of wormwood, the leaves of fennel acquire a succulent structure (Grümmer, 1955, 1957a; Schwär, 1962).

One of the possible general mechanisms of action of cholines is the sensitisation of plants to light, temperature, or any other factor. Thus the choline itself has no visible effect, but the plant has manifold increased or decreased sensitivity to one or other harmful effects. Thus, the effect of a choline, in the absence of an external factor, does not appear, for in itself the choline does not affect the plant.

Returning again to the question of the specificity of action of cholines, it should be noted that it very much depends on a combination of external conditions. For example, in the experiments of S.I. Chernobrivenko (1956), carried out in conditions of the Sinel'nikovskoi Experiment Station, chickpeas was a negative component for corn; in experiments in our laboratory (I.I. Dzyubenko), carried out near Kiev, chickpeas greatly stimulated the growth of corn. V.P. Ivanov (Moscow) in pot experiments, placed one series of vessels with a mixture of corn and beans in the greenhouse, the other in the open air. The results were exactly the opposite. In the greenhouse, corn with beans grew better than in pure sowings. Outdoors, it was sharply inhibited by beans. Thus, the effect of the same concentration of choline may be quite different depending on the overall physiological state of the acceptor plants. Under conditions that are optimal for plant growth, it takes the same concentration of choline to be stimulating, as it does, with a lack of heat, moisture, food, to inhibit growth.

A typical example, that can also serve, is the interaction of pine and birch in mixed plantings. On sites with good forestry conditions, birch displaces pine; in poor conditions, pine displaces birch, and in some mixed plantings with average conditions, they can both exist indefinitely.

It may be that in allelopathy, it is not only the concentration of the active substance, which in bioassay may be in the class of stimulatory, or even inactive, that is important, but it is also the **duration** of exposure, as measured by the period of the life of the plant. The plant is constantly experiencing a negligible effect of cholines entering the environment from the absorbed state, or from any source. What are the consequences that result? This is not yet known. In plant physiology, there are always, more or less, high concentrations at single, short-term exposure. We can assume that the constant exposure to low concentrations of cholines will be "addictive", an adaptation of plants, but on the other hand it may occur and there is the progressive suppression, slowing down of vital functions, and disruption of metabolism. All these issues need to be studied.

For the plant in the ontogenetic cycle there is a period of intensive growth. During these periods, the plant requires the maximal supply of provisions for life, and most strongly reacts to both favorable and adverse conditions in the environment, including the effect of cholines. It is during these periods, the plant responds to the best fertiliser, irrigation, etc. and it is then that it has the lowest resistance to cholines, as at this time the plant is "open" for all sorts of influences. At other times the plant carries on little exchange with the environment, it is poorly responsive to favorable factors and appears to be more tolerant with respect to the allelopathic effects of neighbours.

Such resistance of plants to cholines can be characterised as "passive resistance to transfer by slowing growth." There is also an active resistance when the plant is in a state of intense growth; it overcomes the adverse effects by an even greater increase in metabolism, reorganising it for the destruction of absorbed cholines.

So, plant secretions and metabolic products of heterotrophic organisms have diverse physiological and biochemical effects on plants. Symptoms of these effects are very similar; although the internal mechanisms are not the same, choline action depends on its concentration, on interaction with other substances present in the environment, and the physiological state of the acceptor plant, which is determined by environmental conditions, its age and species characteristics.

Connection of allelopathy with other forms of interaction of plants

The interaction of plants in conjunction with growth is comprised of, on the one hand, using the general sense of abiogenic factors of life - light, water, nutrients, oxygen (in the ground), carbon dioxide (in air), electricity, etc., and, on the other hand, interrelationships that consist of allelopathy, chemical interferences, to which we assign an active character. The latter reflects only the fact that plants cannot actively influence the course of the cosmic and many earthly abiogenic factors, and they are adapted passively to their living surroundings. Only in this sense can one understand the commitment we have made to the division.

In the common abiogenic factors in the life of plants, are ecological conditions that are primary in leading to the formation of plant communities. Even small deviations in the ecology - different exposure of the slope, other micro-relief, changes in the level of ground water, nitrogen, and thousands of these conditions lead to the formation of completely different phytocenoses. Therefore, in the interaction of plants, the sharing, or so-called commensalism, or contestation, competition, the "struggle" for the living conditions plays a major role. By adjusting these conditions, relationships in plant communities can be largely successfully managed, to increase the productivity, durability and resistance of phytocenoses.

However, this does not mean that abiogenic environmental conditions play a solitary role, and that of allelopathy is a small, supporting role. Environmental conditions begin to act at first, they create the foundations, background, on which are based the allelopathic interactions of plants. Other things being equal, the allelopathic characteristics of plants determine the composition of species, intensity of growth, all manifestations of plant life in the community. Therefore, the study and control of abiogenic ecological factors excluding allelopathy may not lead to a complete success.

Moreover, as we have already pointed out, the entire plant life in the community is closely linked with the various parts of the cycle of active substances in the cenosis. The plant community should be seen as a self-regulating complex system composed of many varying components. One of the important means of transmitting information in a cybernetic system is plants, i.e. allelopathy is a channel of direct and feedback links in the system. The community of plants, however young, should be viewed as self-adjusting, not as a random collection of plants, not related by anything, and thus as a natural phenomenon of nature. The emergence of certain communities, their development, change, the prevalence of certain species in the community is a natural phenomenon of nature.

Meanwhile, some geobotanists are on a different path. For example, we quote from the famous textbook by J. Braun-Blanquet (1951): "The division into “accidental”, without benefit to an individual, and "essential" without benefit to individual members or parts of them (Degener) would be impossible for plants. The principle of utility, the division of labour, perceived support, the gathering of all forces to achieve a common goal in the plant kingdom does not exist. Here the struggle for existence dominates without limit; it regulates directly or indirectly, all of the unconscious manifestations of the social life of plants”.

Further, the author divides the relationships into the two main forms: linked to dependence (parasitism, symbiosis), and linked to commensalism (“Le commensal est simplement un compagnon de table” - "Commensalist - a companion at the table, dinner companion"). The latter connection is purely random, arises spontaneously when organisms engage completely independently in a contest and their mutual influence is based only on the fact that they both use a factor of habitat.

This approach to understanding the essence of relationships in the community and patterns of development is metaphysical and leads to a dead end. On the basis of these positions one cannot understand the reasons for the multitude of plant formations, it is meaningless to talk about any laws governing the development of vegetation, which appear in the form of a random chaotic game of uncertainties.

In fact, as we saw above, the plants in a cenosis are closely linked, form a biological chain through the exchange of organic and mineral products. A plant in a cenosis never feeds individually. In crops and natural cenosis, it feeds along with other plants and it times its rhythm of nutrition to the rhythm of the excretory activity of its neighbors (Rakhteenko, 1958, 1963). Root exudates of plants strengthen or weaken the absorption capacity of neighboring plants. Therefore, competition or commensalism in "pure" form, as interpreted by J. Braun-Blanquet and some other authors (Bergh, Elberse, 1962; Welbank, 1961, 1962, 1964; Hull, 1963), does not happen.

The same applies to competition for water. Water balance of plant communities is crucial for their development and productivity. It is no exaggeration to say that in most parts of the globe, rainfall and its distribution throughout the year are the main factors determining the species composition, changing aspects, productivity and durability of plant groupings. Walter (1962) indicates that in areas with an annual total precipitation of 150-1500 mm, the density of the vegetational cover and plant productivity are proportional to the total amount of precipitation.

Competition for water is the main component of the biological relationship between plants in deserts, semi-desert, steppe and forest-steppe zones, and very many crop phytocenoses. Even in areas of normal and excessive moisture, there is often a situation of intense competition for water.

In this regard, there is a question about the role of chemical interaction in the competition of plants for water, the importance of physiologically active excretions in the absorption and expenditure of water by plants. Firstly, one can say that the relationship between these phenomena exists. For example, when drought increases, so does the excretion of amino acids and other substances by the roots. Soil moisture will always contain a certain amount of organic physiologically active substances - waste products of higher plants and microorganisms. It is also known that the effect of phytoncides and water-soluble cholines often leads to wilting of plants, the disruption of their water balance. It can therefore be argued that "pure" competition for water in natural conditional does not occur.

Plant excretions have an effect on photosynthesis, respiration, enzyme activity, etc. Therefore, in other forms of commensalism, the use of carbon dioxide and light, oxygen, reactions to the environment, electropotential and any factors, are associated with allelopathy. Cholines can enhance or retard, shift at other times all manifestations of plant life in the community. In this connection, a plant that finds itself in this or that community, changes the nature of the requirements to the complex of ecological conditions for optimal development. There are so-called autecological and synecological optima in plants frequently sown in a mixture with other species (Rabotnov, 1959).

With regard to the role of allelopathy in relationships of close causal dependency - parasitism and symbiosis, it is also quite definitely here. Since every green plant is capable of heterotrophic nutrition, in principle, it can go, in the process of evolution, to full hemiparasitism and parasitism. This is facilitated by the exchange of organic matter between plants. For these parasites, the presence of secretions from the attacked plants in the environment is important. For example, seeds of broomrape (*Orobanche* L.) may lie many years in the soil, and germinate only under the influence of root exudates of the host plant. Thus, the origin and development of parasitism is based on an allelopathic principle.

At the heart of the symbiotic relationship is the good regulation of parasitism. In case of violation of the symbiotic balance, one of the symbionts here becomes a parasite, and kills its partner. The essence of symbiosis is in the exchange of waste products and very often in immediate chemical interaction by metabolites.

So, allelopathy permeates all forms of interaction between plants. Although, again, abiotic ecological conditions play a primary role in the life of plants and their communities, allelopathy imposes an exceptionally strong imprint on the effect of abiotic environmental factors. We can say that the impact of ecological conditions often is realised through allelopathy.

There are many examples of wonderful plants in some associations and the complete absence of these same plants in adjacent associations, spaced a few meters away, although measurements of temperature, water nutrient and other conditions were unable to establish any significant environmental differences between these habitats. Many plants under artificial sowing are thriving from the Arctic Circle to the equator, but their natural habitats are much more modest in size. There is an impression that the plant as it "learns", "distinguishes between" what is appropriate and inappropriate for its phytocenotic situation.

However, data on the possibility of such "recognition" are almost there. Considering the plant, which is devoid of senses and perceives external influences in all cells, as an organism, it is unable to analyse the causes of these effects, and we must come to the conclusion that the plant is unable to "recognise" its neighbors in the cenosis. It would seem that the plant uses its active allelopathic mode to have effects on other plants diffusively, regardless of the presence or absence of neighbors. With a shortage of water, nutrients or light, the plant develops a bigger root system or stretches toward light independently of the same or another species. Yet such statements obviously do not correspond to reality. In the course of evolution in plants, they have evolved the ability, in a very complicated way, to respond to the environmental conditions in the whole complex of such conditions. The plant as a special form of specialisation of living protoplasm is able to "remember" the whole program of external conditions that are necessary for the process of ontogenesis. Examples taken from the physiology of plants, give it a striking confirmation. In plants, as it has been recorded and transmitted by heredity, there is a program of requirements to external conditions, which is gradually being realised to the extent of the implementation of these conditions in nature. Disruption of conditions or discrepancy from the program can permanently delay the development of plants. This hereditary, recorded program we call the phasic requirements of plants. It arose as a result of adaptation and as a way of adapting in the plant to specific "written in memory"[3] information about the environmental conditions of the plant.

I could cite other examples related to reproduction and dispersal of plants, with the physiology of adaptive movement of plants, etc., which is certainly testimony to the existence of certain mechanisms in plants as "memorisation" in the cybernetic sense defined hereditarily as "memory". In the understanding of these phenomena, we rely only on the achievements of modern biology and cybernetics, and draw no parallels with the "soul of the cell", "cell psyche" of Haeckel. One can assume that plants produce the necessary programs of adapting for allelopathic interactions with other plants, for the "recognition" of neighboring plants, which together have existed together with it for millennia, and the corresponding reaction to their presence. The

---

[3] At first glance, this statement is reminiscent of Lysenkoism, the doctrine of Trofim Lysenko (1898-1976) who held great influence in agriculture and genetics in the USSR, with his distorted views on heredity based on acquired and learned characteristics. During the 1950's, his influence began to wane, and he was finally ousted from his posts in 1965 Grodzinskii was well versed in modern genetical ideas, and had read Darwin early in his career. Any reference to plants "remembering" is his anecdotal way of invoking the concept of adaptation. There are some current suggestions that some plants can modify their growth or reactions following exposure to certain variables, but there is nothing to suggest that this is heritable.

plant receives information from another plant by changing the complex external environment, and first of all, changes the allelopathic environment. This information causes in the plant a certain response, the nature of which depends on many reasons, and is also due to evolution. This information may cause increased growth that we are witnessing in the first invasion of weeds into crops (Laštuvka, 1955b), the delay of germination of seeds of herbaceous plants under the forest canopy (Kozhevnikov, 1950), increased release of a substance (Sabdulaev, Savchenko, 1964), and so forth. The relationship between the generative and vegetative development can change. In this lies, among other things, the reason for juvenility in grassy phytocenoses, when plants, ending in a pure stand, in their cycle of development for two or three years, many years, "sit" in a vegetative state in a cenosis.

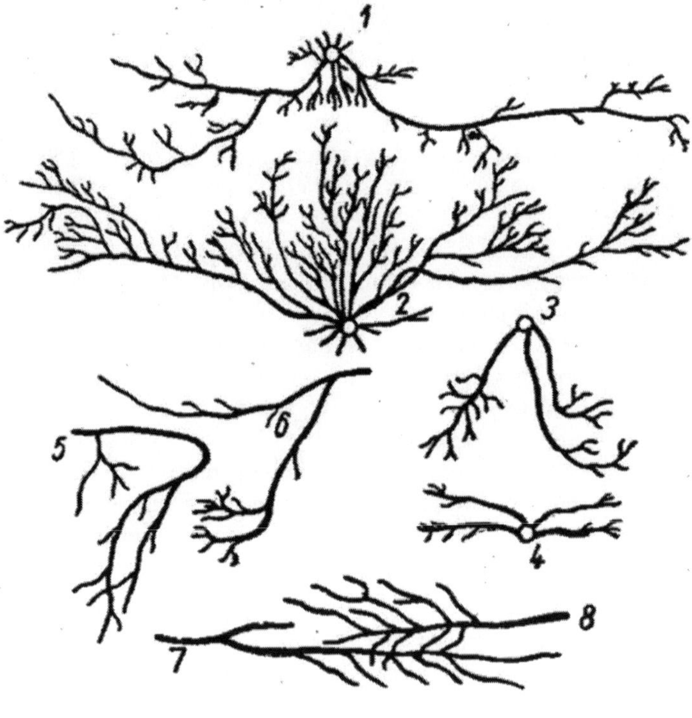

Figure 31. The behaviour of the root systems of fruit trees when placed in the garden: 1-2 – peaches (four years after sowing seed), 3-4 – apples, 5-6 – apples, 7 – apple, 8 - cherry

There are studies that show the positive and negative tropisms roots of various plants to each other (Figure 31).

Of course, to confirm this hypothesis, we need new data.

The use of cybernetics to discuss cooperation in plant community can be very rewarding. Some attempts in this direction were made by V.G. Nesterov and V.D. Alexandrov (1963). In their works are interesting ideas about the role of feedback in the development of plant formations. Unfortunately, the authors made no mention of the chemical interaction of plants.

The influence of cholines, secreted from plants or cenoses, on plant donors or cenosis donor can rightfully be considered the feedback system. Its action can be sustained, and to be a condition of memory, i.e. the ability of a cybernetic system to indefinitely retain the imprint of the influence on the process.

The feedback effects of cholines on generating plant systems are a cause of the self-regulated development of these systems. A small amount of cholines stimulates the growth of biomass in a cenosis, and consequently, the growth of production of new cholines and a feedback reaction; an excess of cholines leads to lower growth. So, the level of cholines in the environment is automatically established.

A choline may be an energetic substance, the effect of which is proportional to the mass, and is a signal in the case when the plants are adapted to react to it, regardless of its concentration. The plant in this case acts as a signal collector.

To sum up this chapter, we can state the importance of organic substances absorbed from the environment in the life of plants. The heterotrophic nutrition of higher plants is a relatively small component - a few percent of the total photosynthetic productivity of carbon, but substances coming into plants can substantially alter their metabolism, growth and quality of the crop. Adjacent growing plants are able to have lively exchange of organic and mineral substances. As phytocenoses develop, so does the general fund of nutrient substances.

A variety of allelopathic, active substances entering the plant, provide multifaceted biochemical and physiological effects. Externally the reaction of an acceptor plant resembles a typical non-specific response of a living organism to stimuli, but the internal mechanism of action to different cholines varies. Cholines alter the permeability of the protoplast, the activity of enzymes, respiration, the absorption of substances and water, and photosynthesis. Cholines of different origins have the potential for physiological interaction, where the secretions are usually found in the nature of paired combinations that reinforce the physiological effect (synergism). When the mixture is administered with a larger number of components, the synergism weakens and is replaced by antagonism.

The influence of cholines depends on the physiological state of the acceptors. Plants in a phase of slower growth or rest are more stable than those that are intensively growing. Under optimal growth conditions, an acceptor receives cholines from another species as a stimulator; and with deterioration in the external environment, the same dose of cholines acts negatively.

Other forms of interaction between plants, the so-called competition or commensalism, are closely intertwined with allelopathy. Chemical secretions of plants are the channels of information and the feedback in the plant community, thus ensuring the self-determination of developing phytocenoses.

Chapter 5

PHYTOCENOTIC IMPORTANCE AND PROBLEMS IN PRACTICAL APPLICATION OF ALLELOPATHY

Even though the problems of allelopathy are only recently gaining recognition and are beginning to be really elucidated, there is still opportunity to provide some actual paths to theoretical development and practical application in conformity with the principles of the chemical interaction of plants.

Without insisting on the absolute reliability of the considerations and hypotheses presented below, we still believe it is necessary to express them, because in any important perspective, it is especially critical for new endeavors.

Phytocenotic role of plant exudates

Many facts, obtained from ecology and geobotany, show that the causes of plant groupings cannot always be explained by the peculiarities of the distribution of vegetative or seed propagules, competition for abiotic factors, or the effect of disparate microclimatic conditions (Garb, 1961). Indeed, in very similar soil and climatic conditions there are often complex associations, and there is a great micropatterning of vegetative cover. In practice, in certain groupings, one never encounters certain species of plants that grow satisfactorily in a number of other, not less dense, phytocenoses. In such cases, it is not abiogenic environmental conditions or competition for life factors that determine the specific nature of the distribution of plants and their combination in phytocenoses. It is only natural that the explanation of these facts involves the chemical interaction of plants.

The secretions of plants have a tremendous impact on the ecological conditions of individual habitats and the surface of the globe as a whole. Suffice it to say that the free atmospheric oxygen formed and maintained at a constant level is due to the vitality of green plants. Absorbing the energy of sunlight and building products from external constituents, plants in general cause oxidation of the surrounding environment, producing a shift of redox potential in the environment toward the side of oxidation, contribute to the accumulation of low-energy products in the environment and high-energy reduced products in their own body.

According to F.W. Went (1962), global vegetation emits 175 million tonnes of essential oils annually. They are oxidised and decomposed when exposed to light, they form black bituminous substances that release 1.75×10^{18} calories a year and cause a permanent positive charge in the atmosphere relative to the earth's surface.

This energy is of the same order of magnitude as the total energy of lightning, calculated by meteorologists to be 1.5×10^{18} calories per year. It is shown statistically that thunderstorms occur mainly in areas with abundant vegetation and are observed only rarely in the polar regions, and over vast deserts and oceans.

Thunderstorms occur very rarely in winter, or early spring and late autumn, i.e., in periods when plants have reduced, or a lack of, secretory activity. Charge distribution in a thundercloud, in the opinion of F.W. Went, is caused by oxidation of organic matter in the peripheral parts (black coloured, positive charge) and the lack of them in the center (white, negative charge). This is the cause of lightning in a storm cloud, and between the cloud and the ground. Perhaps the hypothesis of F.W. Went has not found full confirmation, especially with some particular issues relating to ball lightning or tornadoes, but the argument he holds sounds convincing enough to confirm the idea expressed above, about the great role of plant emissions formed in the environment. For example, it is worth mentioning the accumulation of acidic humus and the formation of podzolic soils in the vast territory occupied by coniferous forests. It is known that the development of these processes is in no small degree due to the excretion and litter of softwoods, and abundant antibacterial substances, that cause predominantly fungal decomposition of organic matter, with the formation of acidic products and the occurrence of podzolisation processes with all the ensuing phytocenotic consequences.

Consider some of the work on the significance of allelopathy in phytocenology. Analysing the causes of poor dissemination of therophytes [1] in rosemary-heath communities in the north of France (in wastelands - garigue) and some other herbaceous communities L. Guyot *et al.* (1951) noted that this was not due to a lack of suitable areas for their growth or seed stocks, because in the designated cenoses, the cover was regarded as only 50% and the number of seeds of therophytes able to germinate was 60 -150 per 1 m^2. The lack of therophytes also cannot be explained by the unfavorable environmental conditions. G. Deleuil (1950), in the same association, observed after strong rain a large number of shoots, which soon died. The soil on which rosemary or heath grew, or water filtered through the soil, acted adversely on the shoots of therophytes, and had no effect on seedlings biennial and perennial plants. This led the authors to assert that the absence of annual plants in the studied cenoses was the result of toxic effects of root exudates from the dominants.

In subsequent papers, the French geobotanists tried to clarify the role of root exudates in the formation of certain plant groups. To get comparable results, the experiments were performed on seeds and seedlings of cultivated plants in Petri dishes with filter paper or sand. They were tested for the effect of whole roots leached with water (1/10), the aqueous extracts of chopped fresh (1/1) and dried (1/4) roots, and the aqueous extract of roots from the soil (1/1), taken from the rhizosphere of the experimental plants. The strength of inhibition of test soil extracts from heath false brome (*Brachypodium pinnatum* (L.) P. B.) – the edificator[2] in this cenosis, was

---

[1] Therophyte is a term coined by the Danish plant ecologist C. Raunkiaer to describe plants whose propagules are principally seeds.

[2] Edificator is a term used in eastern European ecology to denote a plant species that is largely responsible for determining the formation of a plant community of phytocenosis

taken as 100%. This species does not have any toxic effect. All other components of the cenosis markedly retarded seed germination and seedling growth of wheat, flax, radish and vetch.

L. Guyot and M. Massenot (1950) observed that, after the removal of the vegetation cover, top layer of soil and poisonous secretions, the community of heath false brome began to develop different species of plants, including 57 species that previously had been totally absent here. Obviously, there were many seeds in the soil that had germinated, having lingered for the extent of many years. For example, seedlings of common St. John's wort in the three years of observation appeared in these numbers: 1st year- 2000- 4200; 2nd year - 600 – 1100; 3rd year to 500-800 individuals in the plot. In the first year after the removal of the toxic soil, the delay in germination was overcome in most of the seeds.

This makes it possible to reveal the reasons for the weak spread or complete absence of therophytes in some herbaceous phytocenoses.

The second question that occupied L. Guyot and his colleagues was the study of the continuous degradation of grassy stands by mouse-ear hawkweed (*Hieracium pilosella* L.) and other species that form single component associations. By degradation they mean the phenomenon when one type of plant community, temporarily dominates small plots, and after some time, it once again is replaced by other species coming into parts of the community. L. Guyot observed such degradation with mouse-ear hawkweed, yarrow (*Achillea millefolium* L.), wild thyme (*Thymus serpyllum* L.), and species of the genus *Helianthemum* Mill.

The author notes that in stages of the old swards with hawkweed, which can be observed simultaneously, as they are located in concentric circles:

1. On the edge of the patch, there is a narrow belt where hawkweed grows well to give fruition, and is in direct contact with neighboring plants, that also show no abnormalities.
2. Further, inside the patch, there is a zone of varying width where hawkweed development is indifferent and it flowers rarely or not at all. Other plants that are represented here are weak, undersized specimens and bloom rarely, and their leaves are of chlorotic appearance, sometimes with necroses.
3. Still in deeper is an area where the vegetation becomes poorer and hawkweed develops increasingly poorly, and finally disappears.
4. In the center, the surface is completely devoid of vegetation, where seedlings are killed due to poisoning. Only after a long time, is there any new colonisation of the plots by annuals.

In explaining the degradation of the patches by hawkweed, L. Guyot and his colleagues have proceeded from the fact that extracts from the roots of this plant were significantly toxic to many plants. At the same time, the high density of cover is very important, as when the number of plants of hawkweed exceeds a certain value, damage to neighboring plants is observed. Weakening of competition promotes better development of hawkweed, its concentration increases, and finally, hawkweed begins to suffer and is killed by self-poisoning. An important factor for the degradation is the degree of longevity of toxic substances in the soil and the peculiarities of the plant itself. In the above case, hawkweed is represented by many forms, which, obviously, must differ in their patterns of degradation that are not observed uniformly.

The third question in the study that was addressed by L.Guyot and his colleagues (1951), concerns overgrown fallow lands. In particular, the authors ascribe the change from the domination of hemicryptophytes[3] to the predominance of grasses, as due to autointoxication by the former plant by root exudates, similar to the degradation pattern of hawkweed. It is possible to agree with these findings, with only one objection, and that is that it is impossible to attribute the interpretation of the toxic effect due to root exudates. Root exudates are only part of the effect of allelopathic plants and, moreover, are not very active.

Of very great importance is the dead litter in perennial cenoses, and various other organic substances entering the soil. The allelopathic value of litter, until recently was underestimated by researchers, although this issue was raised in the works of S.I. Chernobrivenko (1956) and G. Grümmer (1955, 1957a).

We conducted some determinations on the accumulation of cholines, miasmins and saprolins in various phytocenoses. For analysis were chosen sites of herb-fescue-feather grass steppe, located on plakor [4] and overgrown in places with microassociations of couch grass and pubescent wheatgrass and awnless brome, and areas long left fallow and artificially seeded with awnless brome. The determination of the allelopathic regime in mixed crops of legumes and cereals was performed by N.N. Dzyubenko, and in forested areas by V.K. Myakushko and E.V. Vanzha.

Determination of volatile and water-soluble biologically active substances was carried out with the aboveground parts of plants, the upper partly decomposed layer of litter, the lower layer of litter, the soil of the rooting layer, and the roots or rhizomes, in accordance with the procedures described in the Appendix.

The data from the active extracts or volatile substances, obtained from bioassays were determined using the standard curve for relative units (RCU), and then the content was calculated per 1 m^2 of cenosis for various structural components and in total. Being aware that such characteristics are very largely conditional, we still believe it is possible to consider that allelopathic conditions apply, and are prevailing inside and on the borders of different studied associations.

Let us consider some of the results obtained. Data on the content of cholines in the associations of couch grass (*Agropyrum repens* (L.) R.B) are shown in Table 28 and Figure 32.

The association of couch grass has an area of 105 m^2, and its boundaries with the dominant forb-fescue-*Poa* steppe are not expressed very clearly. Individual runners of the couch grass penetrate 0.5-0.7 m into the neighboring grass stand.

---

[3] Hemicryptophyte is a term introduced by C. Raunkaier to denote a perennial plant whose perennating buds are at ground level, and partly protected by litter

[4] "Plakor" is a Russian term for a type of water-logged plain characteristic of some of Ukraine, Russia.

According to observations of V.V. Osychnyuk in the course of the previous four years, this association invades the mixed herb-*Poa* steppe at a rate of approximately 15-20 cm per year.

In the centre of couch grass patches that were not regenerating couch grass, its growth is weakened. The average height of vegetative shoots in the centre was 66.6 cm. in the zone of maximum development it was 68.3 cm, and at the junction, it was 55.7 cm, Weight of above-ground mass of couch grass for 1 m^2 at these points was 1030.4 g, 1176.8 g and 905.6 g respectively. Analysis of total curves is shown in Figure 32, and the more detailed data in Table 28 show that the total content of cholines, over the 25 days that passed between the two experiments, significantly increased, but the nature of the distribution between the studied groups of cholines was preserved. The highest activity was characteristic of the structural elements of the central part of the patch, the lesser with the zones of optimal growth of couch grass: at the junction of the two groups, total activity slightly increased or remained at the same level and it significantly decreased in the structural elements of herb-fescue-*Poa* group.

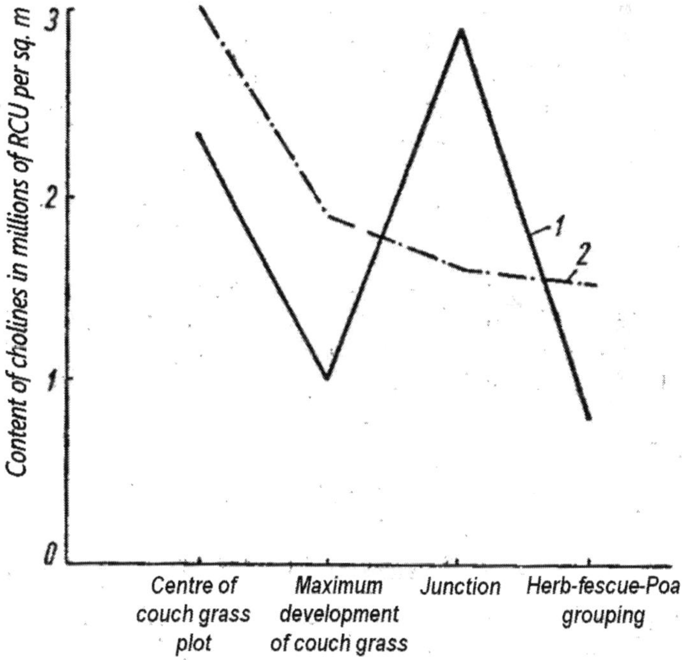

Figure 32. Summary content of cholines in structural elements of two neighbouring associations
1 – 11 June; 2 – 6 July

It is evident, then, that most of the volatile and water-soluble active substances were present in the soil of the cenosis, which contains 68-99% of the total activity. The least active of the total above-ground plant mass was the upper layer of litter. The bottom layer of litter was more active in couch grass patches, but outside of the patches, it was almost equal to the activity of the upper layer of litter. The above-ground mass of roots and rhizomes usually contained the second highest activity, after soil. From this we can conclude that the characteristics of allelopathic stress within a cenosis should suffice for a determination of cholines in the soil. For information about the sources of cholines that enter into the soil, it is necessary in certain cases to analyse other materials.

Table 28. Content of cholines in structural elements of couch grass and mixed herb-fescue-*Poa* groupings (in thousands of RCU per 1 m^2)

Place selected for test	Sample	11. VI. 1963			6. VII. 1963		
		Vola-tiles	Water soluble	Total	Vola-tiles	Water soluble	Total
Centre of couch grass association	Above-ground mass	2.2	17.6	19.8	13.4	69.4	82.8
	Upper litter	0.9	16.3	17.2	1.0	12.7	13.7
	Lower litter	6.5	30.4	36.9	2.0	10.8	12.8
	Soil	1035.0	1242.0	2277.0	873.0	2049.2	2922.2
	Roots	8.0	8.0	16.0	11.0	112.0	123.0
	Total	1052.6	1314.3	2366.9	900.4	2254.1	3154.5
Zone of maximal growth of couch grass	Above-ground mass	7.9	29.4	37.3	15.0	92.6	107.6
	Upper litter	8.3	11.3	19.6	1.7	15.4	17.1
	Lower litter	15.2	23.2	38.4	4.6	78.1	82.7
	Soil	728.0	135.0	873.0	828.0	861.1	1689.1
	Roots	5.0	130.0	135.0	4.0	60.0	64.0
	Total	774.4	318.9	1103.3	853.3	1107.2	1960.5
Junction of two associations	Above-ground mass	4.9	6.8	11.7	3.7	142.2	145.9
	Upper litter	0.7	5.6	6.3	1.6	14.5	16.1
	Lower litter	2.9	2.1	5.0	1.8	16.1	17.9
	Soil	830.0	379.0	1209.0	658.0	773.9	1431.9
	Roots	3.0	62.0	65.0	7.0	73.0	80.0
	Total	841.5	455.5	1297.0	672.1	1019.7	1691.8
Mixed herb-fescue-*Poa* grouping	Above-ground mass	7.9	24.0	31.9	3.0	149.1	152.1
	Upper litter	2.9	6.9	9.8	3.4	6.0	9.4
	Lower litter	6.1	3.8	9.9	1.2	12.2	13.4
	Soil	558.0	270.0	828.0	585.0	779.4	1364.4
	Roots	12.0	26.0	38.0	2.0	66.0	68.0
	Total	586.9	330.7	917.6	594.6	1012.7	1607.3

At the first time, the determination of total activity of volatiles (miasmins, phytoncides) was higher than that of water-soluble saprolins and diffusate from tissues: it changed little over 25 days - in some cases it increased a little, in others it decreased. However, the activity of water-soluble compounds during this period increased by 2 - 5 times and became defining. The increase is observed for all samples, except, perhaps in the roots. This increase can be attributed to the generally observed rise in microbial activity and the beginning of the withering away of the aerial parts, which occurs at the end of June-July-August.

From this follows the second important conclusion that the characteristics of allelopathic relations in cenoses cannot be limited to the single determination. This phenomenon needs to be studied in its dynamics.

In general we can say that couch grass patches have their own higher allelopathic regime, which is 1.5 to 3 times more intense than that in the neighboring steppe. This circumstance facilitates the invasion of steppe by couch grass. However, due to excessive accumulation of cholines, an inhibition of growth begins in the center of the patch, and causes total disappearance of couch grass in the old patches.

Further determinations that we conducted on similar groupings of pubescent wheatgrass (*Agropyrum trichophorum* (Link) Richt.), awnless brome (*Bromus inermis*) Leyss.) located amidst the same steppe, fully confirmed these findings. Each microassociation has a certain level of cholines; the central part of the patch contains a greater amount of active substances than the periphery. In the case of couch grass and awnless brome, there was established quite a clear inverse correlation between the intensity of growth and the content of cholines in the soil. These plants are capable of creating the conditions of autointoxication, self-suppression.

On land long left fallow, the determining of the level of cholines in associations of Austrian wormwood (*Artemisia austriaca* Jacq.), couch grass, small-fruited false flax (*Camelina microcarpa* Anrdz.), meadow brome grass (*Bromus arvensis* L.) showed that the total content of cholines in these newly emerging groups was of about the same order and magnitude as those for the previously described grass-forb associations that have arisen spontaneously among virgin steppe. Attention is drawn to the fact of a sharp increase in total activity at the interface of some weed groups, such as wormwood and brome-false flax, that may be due to synergism between substances contained in the litter and soil.

We have carried out the determination of cholines in the litter and soil under plants of Austrian plants sage (*Salvia austriaca* L.) and under sod, disposed to rows, of fescue (*Festuca sulcata* L.), that appears as the dominant feature of the given cenosis. Results of bioassays in relative units for the samples are presented in Table 29.

These data clearly show a very significant toxicity of soil and litter under rosettes of Austrian sage. This toxicity is due to excretions from all of the organs of sage which also turned out to be much more active in bioassays than those from organs of neighboring fescue plants. It is possible that the accumulation of biologically active substances under Austrian sage causes an almost complete absence of seedlings under its rosette, a depressive effect on neighboring plants and its ability to survive in dense grassy stands. This type of sage has its rosette firmly pressed to the ground, and it has broad, rounded, not particularly dense leaves. These leaves projecting into fescue sod and other plants cause death of part of the sod. The mechanical action of such oppression is not possible. It seems to us that this sage is a typical example of an individual plant that survives in the steppe cenosis because of its allelopathic effects on neighboring species.

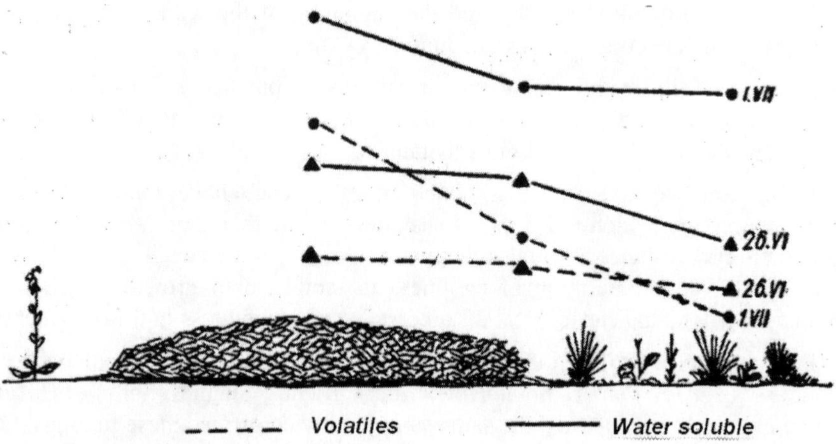

Figure 33. Accumulation of cholines in "cushions" of *Thymus marschallianus*

A similar pattern, but less pronounced, was obtained from determining the water-soluble and volatile cholines in cushions of *Thymus marschallianus* Shilld. In the centre of this cushion (diameter about 50 cm), the content of cholines was approximately 25% higher than outside of it. At the edge of the cushion the content of bioactive substances is almost the same as in the centre, i.e. the cushion supports a well-defined allelopathic regime.

Table 29. Relative activity of cholines under rosettes of Austrian sage and sod of fescue

Object studied	Austrian sage			Fescue		
	Volatile	Water-soluble	Total	Volatile	Water-soluble	Total
Above-ground mass (leaves)	32.7	29.3	62.0	28.6	27.5	56.1
Litter	40.9	45.6	86.5	26.2	13.2	39.4
Soil	71.0	12.5	83.5	22.0	12.3	34.3
Roots	21.2	25.4	46.6	12.0	15.2	27.2

It is curious that the strongly smelling thyme, rich in essential oils, has more active water soluble cholines, whereas the almost odourless Austrian sage has more active volatile excretions.

The accumulation of cholines under individual plants and in their communities occurs not only in the steppe. As shown by the determination of cholines, from oak-hornbeam oak forests in the Kiev, Volyn, and Ivano-Frankivsk regions, as carried by V.K. Myakushko and E.V. Vanzha, in litter, in the soil rhizosphere of oak, hornbeam, cherry trees and other forest species, there is the accumulation of significant amounts of these substances. The content of cholines is greatest in the upper layer of the soil, and decreases with depth: it is higher near the fine absorbing rootlets compared to those that are thick and suberised. The level of cholines gradually increases from spring up to the middle of summer, and then decreases. The accumulation of cholines depends on the soil type. In mixed stands involving two or more species of trees, choline activity is lower than in pure stands.

Woody plants are of particular interest to study the accumulation of cholines, as compared to herbaceous species, they are there for a long time in one place, form a large mass of organic matter and, obviously, create a more stable allelopathic environment. For example, let us examine the effect of planting oak on the growth of corn, sunflower, soy and beans. We obtained data on experimental plots of scientific-experimental facilities of the Institute of Botany, Academy of Sciences of the Ukrainian SSR at Feofaniya. The forest of oak, aged 10 years, is located on the north-east side of the plots with crop plants. The plots were located perpendicular to the forest belt.

Plants located in close proximity to the oaks were extremely suppressed, as the distance from the oaks increased, growth improved; at a distance of 4 to 6 meters, they reached the normal size (as in the rest of the field); in the zone 8 to 10 m from the forest plantations and corn, stimulation of sunflower and soybean growth was observed, within this zone plants were of normal yield.

The inhibition is expressed in plants not only by lesser height before the harvest, but also in their overall size, in the weight of aboveground mass, and lower yields. This can be seen in Figure 34, which shows the soybean plants excavated at intervals of 1 m. The influence of the proximity of the forest plantation on the height of the plant is shown in Figure 35.

Figure 34. Plants of soybean, growing in the vicinity of a forest plantation. Specimens were recorded every meter.

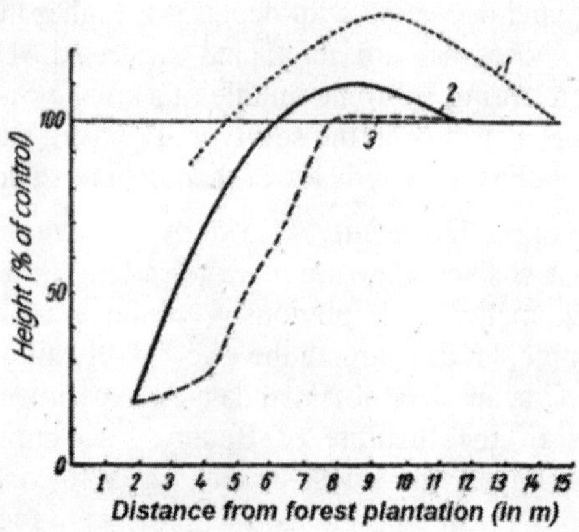

Figure 35. Height of plants grown in the vicinity of forest plantations
1 – sunflower; 2 – corn, 3 - soybean

It is not possible to explain the inhibition of plant growth near forest plantations by the competition for water and light, because the amount of precipitation at Feofaniya in 1962 was abundant. Soil, from spring until the autumn, was almost all the time in water-logged conditions. Lighting conditions for free-standing suppressed plants were considerably better than for non-suppressed plants forming a closed plant stand

One would assume, then, that suppression is caused by competition for nutrients from oak. However, agrochemical analysis of soil from the cultivated layer did not confirm this assumption. As can be seen from Table 30, the potassium content was reduced from the oak forest toward the field, nitrates were detected only in the immediate vicinity of oak, and phosphoric acid content, although it was somewhat lower near the oak than in the field, it was still high enough not to cause such competition between the plants and a sharp decline in growth.

And finally, the fact that stimulation of plants was noted at a distance of 8 to 10 m from the oaks in no way supports competition. Due to this fact it is quite natural to assume a significant role for physiologically active substances released by the oak in these phenomena. We previously established obtaining such substances from oak secretions using bioassays: there are many reports in the literature about this (Kolesnichenko, 1960 b, 1961 a, b).

Such content of inhibitors in aqueous extracts 1:15 from the soil was determined by means of bioassay with radish:

Distance from oaks (in m)	Quantity of cholines (in RCU)
0.5	46
1.8	41
3.2	13
4.6	16
7.4	13
8.1	27
11.6	16

As can be seen, the quantity of extracted soluble inhibitors is quite noticeable near the oaks, and then decreases sharply. More clear results were obtained when rye was grown in soil according to the Neubauer method (Figure 36).

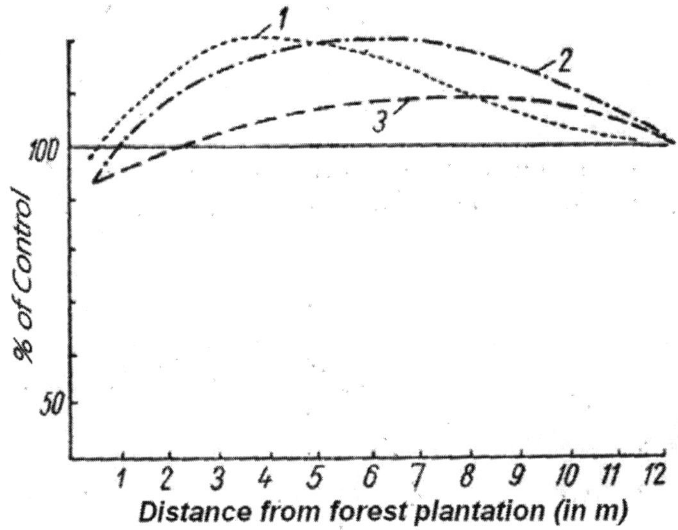

Figure 36. Seedling growth of rye in soil, recorded at different distances from forest plantations
1 – height 2 – weight; 3 - germination

Thus, the above observations allow us to conclude that the protective effect of forest plantations on agricultural crops is to some degree accomplished allelopathically.

Different forest species vary in toxicity relative to the crop plants. Thus, single-row strips of poplar and box elder that grew along the beet fields, caused inhibition of only those plants that were in the immediate vicinity of the poplars. On the other hand, crops in varying degrees, are sensitive to nearby trees. Thus, in the experiment described above, sunflower suffered relatively less than corn and soybeans. In other places, we have observed that tomatoes and potatoes had almost no reduction in growth and productivity in the vicinity of poplar-maple plantings, while soybeans was extremely suppressed, and corn had about three times lower than normal growth, and did not set ears.

It suggests ways to use the influence of forest plantings on the one hand, to incorporate plantings that are harmless to crops, and on the other hand to place alongside plantations crops without reducing yield due to the influence of the secretions of trees.

There is also an accumulation of cholines in annual crops, which was able to be detected by the method of biological assay by N.N. Dzyubenko in our laboratory, and it established a relationship between the content of cholines in the aerial parts, in the soil of the rhizosphere, and the intensity of growth in concomitant cultures.

At an elevated content of cholines there is a noted increase in the components of growth of legume-grass mixtures, and at even higher content of cholines, growth is inhibited. These dependencies can change several times during the growing season, it is still too early to state their value, but they show that every plant, each cenosis forms a particular environment that affects itself and the neighboring plants and the cenosis.

Table 30. Content of mobile forms of nutritive substances in soil near strips of forest (in mg per 100 g of soil)

Height of forest plantation (cm)	Potassium	Nitrate	Phosphoric acid
50	8.4	0.5-1	11.8
180	8.4	trace	11.8
320	8.4	"	15
460	7.0	"	25
740	7.0	"	37.5
810	5.3	"	50
1160	5.3	"	75

In concluding the description of the experiments carried out on this topic, it should be noted that they are still preliminary. When we started these studies, we did not know about the many methodological difficulties that were to arise in the course of the work itself. Facts that were discovered forced some change to the approach of the study of the accumulation of cholines in the structural elements of cenoses. The content of cholines proved to be susceptible to fluctuations that occurred sometimes within a day or two. Thus, the content of cholines in the steppe litter, immediately after rain quickly decreases, apparently due to its being leached, but after a day or two, it sharply increases and significantly exceeds the level that was before the rain. The accumulation of cholines in the environment is subject to slower fluctuations associated with seasonal changes in temperature, precipitation, and so on. Therefore, the characterisation of the allelopathic regime of a cenoses and its individual components cannot be done with a single definition. It is necessary to study the dynamics of changes of cholines in the different structural elements of various cenoses during the growing season to find the dependence of these changes on environmental conditions.

In discussing our data and those from the literature, it is noticed that some plants – edificators, for example *Brachypodium pinnatum*, fescue, feathergrass and other grasses, which occupy fallow land in the later stages in the formation of steppe, and then the main tree species and other plants that make up the bulk of phytocenosis, are not particularly toxic to other plants.

Other species, such as couch grass, brome, and hawkweed and so forth, that form the short-lived "pure" association, are more active in allelopathic relations. Finally, some singly occurring plants, types of "tumbleweed", or even some annuals have very noticeable phytotoxicity. This gives us the opportunity to divide the existing plant species into three groups according to their capacity of having an allelopathic impact on the environment and their importance in associations as dominants or edificators.

The first of these are plants with very active excretions. Owing to these features, they are able to relatively easily penetrate into the dense swards of other species; however, they cannot survive for long in one place, and in one or two years, they disappear due to autointoxication. These plants do not occurs as dominants, or only temporarily dominate for a short time in some groupings (for example, in the first stages of tall weed fallow).

The second group consists of plants with relatively high toxicity, and in the majority of cases, with a vegetative habit of reproduction, so that they form clumps, groves of almost pure stands of a given species. They can dominate a cenosis for several years to decades or centuries, if the environmental conditions prevent the accumulation of cholines above a level harmful to them. Through such allelopathic properties, these plants are well established competitors, but sometimes vanish from the scene due to autointoxication.

Finally, there is the third group of allelopathically weak plants that includes most of the widely distributed dominants. Due to the low toxicity of their excretions, they can indefinitely occupy the land. Their allelopathic environment does not prevent the existence of adapted species and appears to be unsuitable for non-adapted species. Being the most important in a cenosis due to their mass, they create a stable allelopathic background.

These three groups of plants are consistently encountered together. The share of each group in separate phytocenoses varies. In the first stages of colonisation of soil, there is prevalence of allelopathically stronger species, they are replaced by plants of average toxicity, and finally by the weakest.

Practical aspects of allelopathy

As we have seen above, many empirically developed methods of farming unconsciously take into account the peculiarities of the chemical interaction of plants. Of course, in knowing the actual significance of certain effects, one can build a much more efficient system of agricultural practices.

In the problem of allelopathy, from a practical point of view, we must distinguish mainly two aspects: the influence of physiologically active secretions on plant growth and the amount of the useful part of the harvest, and the effects of these secretions on the quality of plant products. In addition, a number of other important circumstances should be considered: the role of secretions in the progressive improvement of soil fertility; its role of excretions in the formation of the soil microflora, including those that are phytopathogenic, in repelling or attracting harmful insects, pollinators, etc.

The main objective of the first of these trends is to master the methods of determining cholines in the environment, to research the stability and requirements of different crop plants with respect to cholines, and the development of measures to regulate the level of cholines in the soil and air. We already know that the optimal content of cholines of plants can significantly accelerate their growth, and all things being equal, it is better to use available conditions to produce higher yields.

The study of the accumulation of cholines in the soil is very important from the point of view of soil sickness, crop rotation, and especially during the intensive use of the soil, for example in greenhouses and hothouses. This issue is extremely important in hydroponics. In particular, the root exudates, and especially the root residues of the previous crop adversely affect the growth of plants, and thus frequent change of the nutrient solutions and the substrates is required. The use of interim and post-harvest crops in many areas may be limited by the presence of highly active cholines derived from excretions, from dead root system, and post-harvest residues of the main crop. The development of agrotechnics with intensive cropping must reckon with this fact.

The impact of allelopathy is very important in pure and mixed crops. The thinning of denser plantings is unlikely to be a useful method, as the remaining roots rotting in the soil can poison the rest of the plants. Thinning should be avoided by sowing at first the required number of plants. As is known, additional sowing or replanting of young plants among more adults in cereal stands almost always ends in failure, due to the allelopathic influence of stronger plants on the weaker. Consequently, the ratio between the components of seeding should be chosen carefully taking into account the possible allelopathic effects, and so that all the valuable components develop in the best way. For example, when corn and peas are sown simultaneously, the latter comes up earlier and stunts the corn with its secretions. The foremost farmers in the Poltava region sow peas later, when the corn has begun to germinate. In this case, there proves to be a balance of power of equal partners, and both develop to a good size.

Knowledge of conformity to the principles of the chemical interaction of plants can be very useful to combat weeds, pests and diseases of plants. Although synthetic chemicals are becoming more widely used to combat pests and diseases of crops and for the destruction of weeds, the extent of their use is limited, on one hand, by economic considerations, and on the other hand by the fact that the substances used are often harmful to humans and animals, and their accumulation in the soil and the crop can lead to undesirable consequences. Because of the genetic instability of pests and pathogens, chemical means of control do not always retain dependable effectiveness. On the other hand, rapid chemical intervention in the established natural relations can lead to unexpected and unwanted destruction of useful plants and animals or to the steady disturbance of the balance in nature.

Measures for deterring harmful insects or preventing fungal and bacterial infections, based on the use of physiologically active plant secretions, devoid of the above drawbacks, should be developed and applied in conjunction with the use of pesticides. Of great practical interest is the investigation into the phytoncides of crop plants and the associated resistance of plants to diseases and pests. (Tokin, 1960). In certain cases phytoncides, act as repellent substances, warding off insects from neighboring, affected plants (Berderevskii, 1961).

In a system of measures to combat weeds, one should introduce "allelopathic weeding", which consists of using allelopathic antagonists against certain weeds. For example, according to A.L. Kursanov, in the Netherlands, in reclaiming the drained seabeds with crops, the planting of wheat was used successfully to control reeds. The inhibitory effect of rye on a number of weeds has long been known. In particular, rye can significantly reduce the infestation of fields by wild oats (Aamisepp, Osvald 1962). Herein are possible and provocative methods. For example, the sowing of oats dramatically stimulates the waking of wild mustard seeds in the soil. If you mow the oats for green fodder before contamination by mustard, it can be seen to reduce seed stocks of this weed in the soil for the occupied fallow soil.

In addition to the active vital emissions and naturally dead parts of plant, enrichment of the environment of the cenosis with organic matter, in crop cenoses an important source of cholines occurs through ploughing the soil with the crop residues, weeds, green manure crops and organic fertilisers. This means of enriching the soil with organic matter is significantly different from the processes in natural communities.

It is important to note that if in the ancient system of agriculture man has sought to reduce the level of organic matter to speed up mineralisation, then with the intensification of the agriculture and the "ploughing up" of soil increasing, there is a need for additional input of the organic matter to the soil. Clearly, the optimal growth of plants needs a certain amount of mobile organic matter in the soil. Here the quality of this matter is important from the allelopathic point of view.

It is known that in the harvested field up to 50% remains in the form of root residues, stubble, leaves, straw and chaff; all 100% of the mass is retained in ploughed soil, in the case of weeds or crops cultivated for green manure. The question arises, whether these materials, containing substances that are harmful, inhibitory to growth, can reduce crop production or impair the subsequent crops?

On the basis of published data, this question can be answered in the affirmative. Some examples of the negative impact of crop residues are in the works of S.I. Chernobrivenko (1956) and G. Grümmer (1955, 1957a). In a series of studies from a group of German authors (Winter, Bublitz, 1953a; Winter, Schönbeck, 1953; Winter, Sievers, 1952, Bublitz, 1953), it has been shown that aqueous extracts of fallen leaves of beech and pine, or straws of cereals inhibit seed germination and growth of watercress, wheat, poppy, oats and other plants. In recent work by E. Sievers (1961, 1963), he demonstrated the presence of antiphytotic substances in vine shoots, which, after pruning got into the soil. These substances retarded germination and growth not only of lettuce and wheat, but also vines, i.e., they are not specific.

As shown by W. Flaig and E. Saalbach (1959), the introduction into soil of very small amounts of straw significantly enhances the growth of young plants of rye and their absorption of nutrient substances. These are explained by action of the physiologically active substances produced by the decay of the straw.

A literature review (Börner, 1960) lists the facts of adverse impact of the root residues of peach, apple and citrus, which prevent the regrowth of horticultural plantations. The harmful effects of decomposing roots of peach are determined by enzymatic degradation of products of amygdalin (prussic acid and benzaldehyde) and in apple, by the release of phlorizin and its derivatives.

However, a systematic study of the spread of inhibitory substances in the post-harvest crop residues has not been conducted. At the same time, it is of practical importance for the development of agronomic practices that would eliminate the harmful effects of inhibitors.

Likewise, hardly studied is the spread of the cholines in the remains of weeds, that during ploughing, cultivation and weeding get into the soil, and can have harmful effect on seed germination and plant growth.

Practical observations show that the introduction of straw or manure on sandy soils leads to the burning of plants; outward signs of such burning resemble symptoms of plant poisoning by physiologically active substances. Experienced gardeners recommend removing the remnants of pulled weeds, because leaving them near plants harms the crop. V.F. Grodzinskaya[5] observed that destruction and incorporation into the soil of the aerial parts of yellow foxtail was more damaging to millet, than maintaining the weed.

In order to verify the possibility of similar activities of weeds residues on the growth of agricultural crops, we have set up a plant experiment, in which 4 g lots of fresh weight of weeds was added to 1 kg of soil (chernozem) at a depth of 5 cm, for the sowing of sunflower. The effect of weed residues on the initial growth of the plants can be seen in Figure 37.

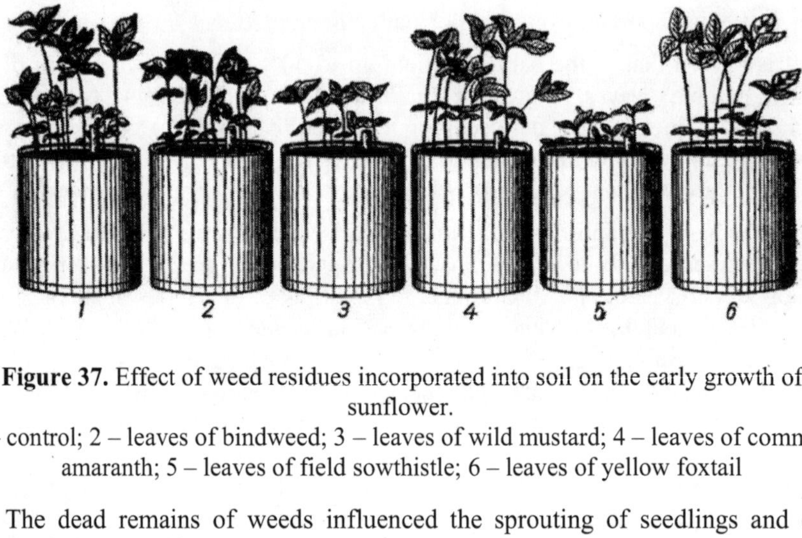

Figure 37. Effect of weed residues incorporated into soil on the early growth of sunflower.
1 – control; 2 – leaves of bindweed; 3 – leaves of wild mustard; 4 – leaves of common amaranth; 5 – leaves of field sowthistle; 6 – leaves of yellow foxtail

The dead remains of weeds influenced the sprouting of seedlings and early growth of plants (Table 31).

All weeds except yellow foxtail, inhibited sprouting of seedlings and the growth of plants. This inhibition decreases with time, but the differences between the samples were maintained throughout the experiment.

A similar pattern was observed in other pot experiments. Sometimes growth retardation was replaced after three to four weeks by faster growth, so that the test plants caught up with the control plants, or even surpassed them. Thus, the various residues that enter into the soil, may significantly affect in some measure the initial growth of plants.

---

[5] An interesting inclusion in the book, as this is Grodzinskii's mother.

Table 31. Influence of weed residues on germination and growth of sunflower

Experimental sample	Appearance of germination (% of control)	Height of plants at 14 days (in cm)
Control	100.0	8.0
Field bindweed	81.0	6.4
Wild mustard	69.2	5.8
Common amaranth	96.1	6.6
Field sowthistle	42.3	5.3
Yellow foxtail	100.0	8.9
Fat hen	65.4	6.2

For a systematic study of the accumulation of cholines in crop residues and in weeds, we used the method of biological assays. For the analysis were selected samples of crop residues of agricultural crops and weeds during July-October 1961 in the fields of scientific and experimental base of the Institute of Botany AN USSR (Feofaniya), Regional Research Station for Animal Husbandry "Terezino", educational sector Bilotserkivskyi Agricultural Institute, Bilotserkovskyi Experimental Breeding Station and Sugar Beet Farm named after "9 January" (Belotserkovsky Kiev region). Samples were dried to air dryness and analysed, volatiles were studied from freshly harvested material.

Our task included biological determination of inhibitors, so when analysing the extracts, we were limited to only a certain pH and the presence of tannins, amino acids, sugars and substances that were fluorescent or UV-absorbing.

We studied the water-soluble inhibitors in 128 samples of weeds and 88 samples of stubble from crops. Results processed by the method of statistical variation are shown in Figure 38.

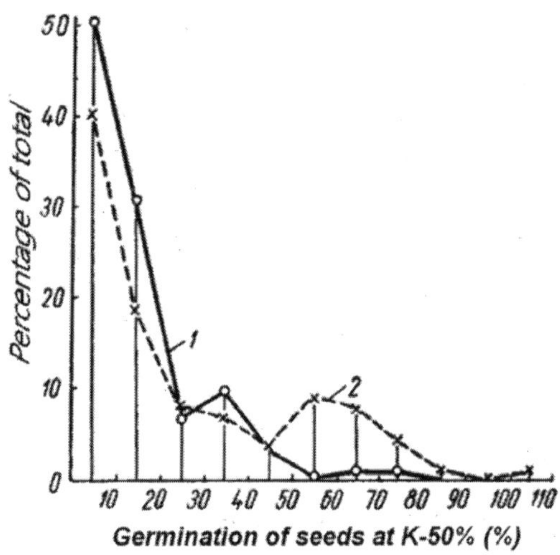

Figure 38. Prevalence of inhibitors in the residues of weeds (1) and in residues of crop plants (2)

From these graphs, one can see a very large distribution of inhibitors from plant materials entering the soil. From the studied group, weeds had the highest activity, where germination in bioassay with extracts that was only 0 - 10% of the control in water, occurred in half of the samples; from the group of crop residues such activity was found in 40% of the samples. No inactive species were found among the weeds; among samples of crop residues, only 2% had no effect on seed germination. Thus, with the soil, in all cases where the crop residues or above ground mass of weeds remain in the field and is ploughed into the soil, it is necessary to consider the presence of inhibitory substances in them.

The sensitivity of seeds of radish, poppy, wheat, lettuce and two varieties of rye to water-soluble cholines tested by us was different. Radish, poppy and salad are the most sensitive and very clearly respond to the presence of inhibitors in the solution. Wheat and rye react more weakly and do not indicate low concentrations of inhibitory substances. However, the overall reaction in almost all cases is quite similar in nature, and therefore for the detection of cholines any seeds can be used. Interestingly, radishes respond well in their germination, but the subsequent growth of radish seedlings depends very little on inhibitors. Wheat behaves differently; there is very little reduction of germination, but it responds sharply to cholines in seedling stage.

Among the residues of cultivated crops, those particularly rich in inhibitors are the tops, the mass of lupine and phacelia for green manure, the tops of potatoes, tomatoes, sweet and hot pepper, cabbage leaves, carrots, parsley, horseradish, beans, sunflower (700- 1200 RCU). There is an average activity of 300-700 RCU in stubble of perennial herbage (clover and alfalfa), Sudan grass, vetch-oat mixture, and corn for green fodder, and hemp leaves. Relatively weak activity is characterized by the remains of cereals, legumes and cereals harvested in a state of full ripeness - stubble and chaff of wheat, rye, oats, barley, millet, buckwheat, and broad beans.

Fresh green leaves contain quite a lot of cholines. With ageing and yellowing, the amount of cholines extractable with water increases, and after death, it is reduced, apparently due to leaching of residues. For example, cabbage leaves, collected on one day from one plant, had the activity (in RCU): green - 760, yellow – 1100, and dead, dry leaves - 367.

Sunflower leaves in the same experiment had: young - 750, old - 835 RCU. Fresh horseradish leaves collected on August 30 had activity of 460 RCU, yellowed leaves, collected October 19 - 1280 RCU. Sugar beet green leaves yielded an extract of 1220 RCU, and already dead, dry leaves - 1100. Stems typically contain less inhibitory substance than the leaves. While in the field, the quantity of inhibitors in the stubble is gradually reduced. We analysed samples of crops residues collected from the same field with an interval of 15-30 days. The content of cholines of the stubble of broad beans over 37 days decreased from 204 to 42 RCU, winter wheat over 12 days - from 88 to 64 RCU

During the decomposition of straw or stubble, the amount of cholines extractable with water appears at first to increase and only then to decrease gradually. Thus, a few samples of fresh pea straw had activity of 60-124 RCU, rotting 180 RCU. The loss of activity occurs generally fairly slowly. Thus freshly harvested stubble of millet had an activity of 50 RCU and rotting straw of millet, exposed for a year to the open air, still contained 31 RCU.

Plant roots contain the least amount of cholines. For example, leaves and stems of potato had activity of 401 RCU, roots - 245 RCU; in beans, respectively - 495 and 31 RCU.

The distribution of the water-soluble inhibitors of germination from the residues of weeds has such features. The highest inhibitory activity was distinguished in extracts of white sweetclover (*Melilotus albus* (L.) Desr.), fumitory (*Fumaria vaillantii* Lois.), wild mustard (*Sinapis arvensis* L.), barnyard grass (*Echinochloa crus-galli* (L.) Roem. *et* Schult.), couch grass (*Elytrigia repens* (L.) P.B.) gallant soldier (*Galinsoga parviflora* Cav.), cypress spurge and sun spurge (*Euphorbia cyparissias* L.; *Eu. helioscopia* L.), hedge-nettle betony and marsh betony (*Stachys annua* L., *S. palustris* L.), barbary box-thorn (*Lycium barbatum* L.), peppergrass (*Lepidium ruderale* L.), cocklebur (*Xanthium strumarium* L.) dandelion (*Taraxacum officinale* Wigg.). mallow (*Malva neglecta* Wallr.), field pennycress (*Thlaspi arvense* L.) – over 1000 RCU.

There is somewhat lower activity in horsetail (*Equisetum arvense* L.), thorn-apple (*Datura stramonium* L.), stinging nettle (*Urtica dioica* L.), a number of dodders (*Cuscuta* sp.), marsh elder (*Iva xanthifolia* Nutt.), common toadflax (*Linaria vulgaris* Mill.), tripartite beggarticks (*Bidens tripartitus* L.) field sowthistle (*Sonchus arvensis* L.), common chickweed (*Stellaria media* (L.), white amaranth (*Amaranthus albus* L.), broadleaf campion (*Silene latifolia* (Mill.) Rendle *et* Britt.) and many others.

Medium strength inhibitors are contained in the green mass of common groundsel (*Senecio vulgaris* L.), black nightshade (*Solanum nigrum* L.), goose cinquefoil (*Potentilla anserine* L.), annual knawel (*Scleranthus annuus* L.), species of orache (*Atriplex* sp.) and goosefoot (*Chenopodium* sp.), creeping thistle (*Cirsium arvense* (L.) Scop.), spiny cocklebur (*Xanthium spinosum* L.), field mustard (*Brassica campestris* L.)[6], scentless mayweed (*Chamaemellum inodorum* (L.) Vis.)[7], yellow foxtail and green foxtail (*Setaria glauca* (L.) P.B., *S. viridis* (L.) P.B.), Canadian fleabane (*Erigeron canadensis* L.), sweet wormwood and mugwort (*Artemisia annua* L., *A. absinthium* L.) and others.

Lowest activity in our experiments, was identified from extracts of common amaranth and slender amaranth (*Amaranthus retroflexus* L., *A. blitum* L.) whorled foxtail (*Setaria verticillata* (L.) P.B.), black henbane (*Hyoscyamus niger* L.), common knotweed, copse bindweed, dragonwort (*Polygonum lapatifolia* L.[8], *P. dumetorum*[9] L. *P. bistorta* L.)[10], field bindweed (*Convolvulus arvensis* L.), blue cornflower (*Centaurea cyanus* L.).

---

[6] Now regarded as *B. rapa.*

[7] Now known as *Tripleurospermum inodorum*

[8] Now regarded as *P. aviculare*

[9] Now known as *Fallopia dumetorum*

[10] Now known as *Persicaria bistorta*

As in the case of other types of substances, the content of the water-soluble inhibitors of the aerial parts of weeds in other groups of plants far exceeds their content in roots (Table 32). Young plants contain more inhibitory substances than old plants. For example, rosettes of common thistle (*Onopordon acanthium* L.) contain 840, but adult specimens 323 RCU: shepherd's purse (*Capsella bursa-pastoris* L.) as rosettes have 1100, but in the flowering specimens, 890 RCU.

The nature of inhibitory action, obviously, is not associated with a change in the reaction of environment, because the pH test solutions are primarily held within 5.0 - 6.5. It is more likely that in some cases, the inhibition of germination could invoke tannin substances, which are frequently found in the water extracts. In the experiments described here, the tannins have been found in significant quantities in 60% of samples of the weeds and in 35% of the samples of crop residues.

Table 32. Content of cholines in extracts from above-ground and below-ground parts of weeds (in relative units)

Plant species	Roots	Above-ground parts
Yellow foxtail (*Setaria glauca* (L.) P.B.)	127	262
Barnyard grass (*Echinochloa crus-galli* (L.) Roem. *et* Schult.)	29	420
Canadian fleabane (*Erigeron canadensis* L.)	121	424
Fat hen (*Chenopodium album* L.)	45	398

However, the highest correlation with inhibition was with the presence of amino compounds, which give a color reaction with ninhydrin. It can be noted that the most inhibitory activity is shown, as a rule, by young, protein-rich parts of plants . Therefore, we can assume that the products of the degradation of proteins account for a significant portion of cholines that we found.

There are relatively few luminescent substances in the extracts, but fairly often these are found to absorb ultraviolet light. The latter, in particular, are degradation products of nucleic acids, having a high physiological activity, and this confirms the assumption made above about the involvement of protein autolysates in the inhibition of growth.

At the same time, there are the so-called substances of secondary origin - organic acids, tannins, various pigments, and so forth, which are often found in extracts. Clearly, these substances may be inhibitors of germination and growth. Unlike the products of protein hydrolysis, which are common to all plants, these compounds are of a strictly specific nature.

Remnants of weeds, destroyed during cultivation, may have an effect on crop plants also because of their volatile emissions. The activity of phytoncides in bioassay is shown in Table 33.

Black nightshade (*Solanum nigrum* L.), wild lettuce (*Lactuca scariola* L.), great burdock (*Arctium lappa* L.), yarrow (*Achillea millefolium* L.), dove's-foot cranesbill (*Geranium molle* L.), barnyardgrass (*Echinochloa crus-galli* (L.) Roem. *et* Schult.), field sow thistle (*Cirsium arvensis* L.) delayed the germination of radish and poppy by 10-20%: common amaranth (*Amaranus retroflehus* L.) and common groundsel

(*Senecio vulgaris* L.) were inactive, while the macerated mass of a couple of species, corn poppy (*Papaver rhoeas* L.) and mugwort (*Artemisa vulgaris* L.), stimulated germination of 6-10%.

Table 33. Influence of volatile excretions of weeds on germination of seeds (germination of radish and poppy at K-50%)

Plant species	Stage of plant	Radish	Poppy
Stinking pigweed (*Chenopodium foetidum* L.)[11]	Flowering	5.5	0.0
Marsh elder (*Iva xanthifolia* Nutt.)	Maturing fruits	–	10.0
Elscholtzia patrini	Maturing fruits	29.8	–
Canadian fleabane (*Erigeron canadensis* L.)	Flowering	51.5	52.4
Bitter wormwood (*Artemisia absinthium* L.)	Maturing fruits	61.0	55.0
Couch grass (*Elytrigia repens* (L.) P.B.)	Maturing fruits	63.0	73.8
Portulaca (*Portulaca* sp.)	Flowering and maturing	65.0	97.6
Field bindweed (*Convolvulus arvensis* L.)	Leaves dying	71.0	87.6
Fat hen (*Chenopodium album* L.)	Maturing seeds	72.3	70.0
Knotgrass (*Polygonum aviculare* L.)	Flowering and maturing	75.0	91.2

From these data obtained on material collected on October 6, when the activity of the vital processes had already died down, it can be seen that more that half of the studied species display a significant negative effect on seed germination. Some of the plants at this time were inactive; some accelerated seed germination. It can be seen that the volatile phytotoxic substances are not only from fragrant, aromatic plants, but they are also from plants, not characterised by a special smell. The influence of phytoncides on the germination of radish and poppy in many cases is almost the same, indicating the general nature of these toxic substances. Similar sensitivity was exhibited in seeds of lettuce, wheat, yet they almost do not react to volatile excretions.

With the aging of the plant, the activity of volatile substances becomes weaker. Even though collected at the one time, the younger weeds had 1.5-10 times more volatile production in terms of relative units than older plants.

The most active phytoncides in our experiments have been shown in plants collected in June - July, during the maximal intensity of growth.

A powerful source of cholines in agrocenoses can be also the above-ground mass of weeds that have fallen into the soil during cultivation or ploughing, the crop residues, and various field, garden, meadow plants that are not used on the farm. As we have noted above, the method of biological test, in spite of the great advantages, has one drawback - its data do not always satisfactorily reflect the relationships that occur in nature.

This happens because bioassay is conducted for a very short time, on discrete biological material, to the exclusion of the majority of factors acting in

[11] Now known as *Dysphania schraderiana*

biogeocenoses - soil microorganisms, other plants and excretions of their substances, etc. Substances active in assay, can be inactive in cenoses due to losses in the air and in the soil, rapid degradation by microorganisms, soil absorption, interactions with other substances, or due to the fact that adult plants respond differently than those in bioassay, or simply because in the wild they cannot have an impact.

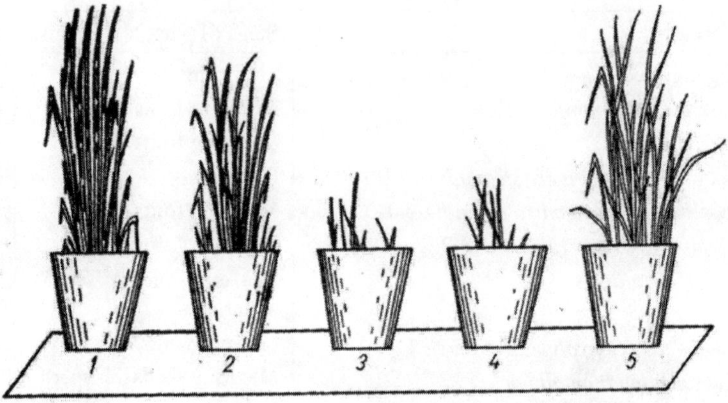

Figure 39. Effect of cholines from Tatarian bread on the growth of oat in sand culture. 1 – control; 2 – mulch from leaves of Crambe; 3 – 1% leaves of Crambe; 4 – as before, 0.5%; 5 – as before, 0.1%

Since these and many other factors, difficult to take into account, contribute largely to reducing the concentration of cholines, one would expect to find that phytotoxic substances will, on the contrary, become stimulatory or completely lose their activity.

The following was carried out in sand culture in pots (2 kg of sand) with six replicates. The aim of the experiment was to check the influence of cholines from Tatarian bread (*Crambe tataria* Sebeök.) on whole plants with exclusion of the influence of the soil. The test results are shown in Figure 39.

As can be seen, oats proved to be very sensitive to the action of the inhibitors from the leaves of Tatarian bread. The energy of germination and even plant growth was decreased by adding 0.1% of the powder of leaves to the sand. With the application of mulch or larger additions, the suppression was even greater. Shoots had a suppressed look, were of bluish color; leaves after unfolding were limp, sometimes with a reddish pigmentation.

The weight of the plant at the end of the experiment was slightly higher with 0.1% addition of the powder to the sand than in the control, i.e. as was said here already, there is a stimulating effect of cholines of *Crambe*.

In the second experiment, we studied the influence of residues of the aerial parts of black henbane (*Hyoscyamus niger* L.) and thorn-apple (*Datura stramonium* L.) on the emergence of seedlings and growth of wheat in sand culture. In the vessels, there was 0.4 and 1.6% air dry weight material, coarsely ground, relative to the weight of sand. The dynamics of the germination and growth in height of plants are shown in Figures 40 and 41, in which it can be seen that the lower concentration of the residual material enhances the growth of wheat seedlings. At high concentrations, there occurred a strong inhibition of germination and growth of the aerial parts. The roots were very much damaged, browned and perished at both application rates of residues in the sand.

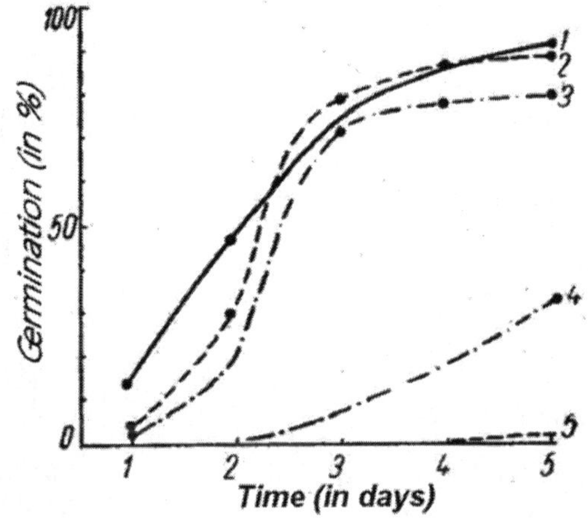

Figure 40. Effect of residues of henbane and thorn-apple on the emergence of shoots of wheat in sand culture.
1 – control; 2 – 0.4% thorn-apple; 3 – 0.4% henbane; 4 – 1.6% henbane; 5 – 1.6% thorn-apple

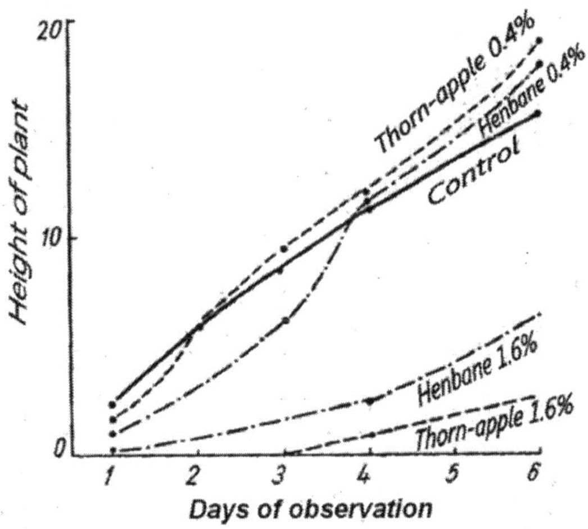

Figure 41. Effect of residues of henbane and thorn-apple on growth of wheat in sand culture.

Thus, these and some other similar experiments in sand cultures showed that cholines detected with *in vitro* methods of biological tests on the seeds, coleoptile segments, cuttings of leaves, etc., act in a similar way to that on whole plants, if the influence of soil is eliminated.

In the experiments with soil cultures, as might be expected, the effect of cholines on plant growth was significantly weakened, and in most cases was positive. For example, in one experiment, several plant species planted in clay pots with sod-podzolic soil, were watered with 1% extract from the fruit of Tatarian bread. The extract was prepared by steeping dry fruitlets portions 100-fold amount of tap water. Control vessels were watered with tap water.

Watering with extract did not significantly affect seedling emergence of yellow lupine, barley and bromegrass, and delayed the germination of lettuce and accelerated the germination of radish.

The results of the effects of extracts on the growth of the aerial parts of the test plants at 28 days after sowing are visible in Figures 42 to 45.

Figure 42. Effect of irrigation with 1% extract from fruits of Tatarian bread on barley. 1 – 2 – control; 3 – 4 – experimental plants.

Figure 43. Effect of irrigation with 1% extract from fruits of Tatarian bread on radish. 1 – 2 – control; 3 – 4 – experimental plants.

Figure 44. Effect of irrigation with 1% extract from fruits of Tatarian bread on lettuce. 1 – 2 – control; 3 – 4 – experimental plants.

Figure 45. Effect of irrigation with 1% extract from fruits of Tatarian bread on awnless brome.
1 – 2 – control; 3 – 4 – experimental plants.

This stimulation concerns, basically, only above-ground organs; roots were the same in radishes, barley, brome grass; were the same size as that of the control; and were only somewhat larger in lettuce.

The experimental plants had a more intense dark green color and developed a larger leaf surface. Despite the fact, that the irrigation water carried a significant number of inhibitors, sod - podzolic soil inactivated their toxic effects, in maintaining their concentration at a level healthy for the plant.

We found a very similar, but more pronounced effect of stimulation, in introducing a powder of the leaves of Tatarian bread to the surface of the vessels with yellow foxtail (*Setaria glauca* (L.) P.B.) and other plants in the same sod-podzolic soil, which can be seen on Figure 46.

Leaves of the plant that came in contact with the powder, were stricken and killed, but in general the plants developed very luxuriantly, several times greater than the control.

In other experiments, when introducing 0.4% residues of certain weeds to the soil in vessels (by weight of the raw soil material) we normally could note the decrease in germination in oats, sunflower, corn, and more or less slowing their growth, which not infrequently is followed by stimulation, and such that in the end the experimental samples were close to the control. This is probably due to the decomposition of cholines and partly to the adaptation of the plants. The residues of some plants, such as alfalfa, caused a very stable and long-lasting inhibition; with other plants it passes quickly.

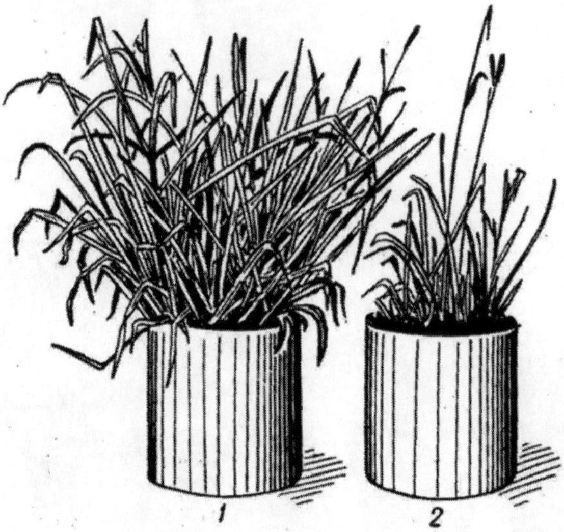

Figure 46. Effect of powdered leaves of Tatarian bread on yellow foxtail
1 – experimental plants; 2 - control

This amount - 0.4% raw mass of the weed weight in relation to the soil - was calculated by us for the arable horizon in relatively average infestations. In the case of larger development of weeds and ploughing them into the soil, the toxic effect increases accordingly. With this, in reckoning with the conditions of low-humus soils, there is a need to be aware of the incorporation of various residues into the soil just prior to planting crops.

Thus, in spite of the buffering role of soil, of increasing the concentration of inhibitors, in many cases it is possible to obtain the same results as in bioassays. Whole plants react as do their parts or germinating seeds in bioassays. However, depending on the point of application, the reaction of various organs is mixed. Thus, with inhibitors entering into the soil, roots are suppressed most strongly and even die, while the aerial organs grow better. Roots generally are one to two orders of magnitude more sensitive to cholines than the leaves.

Breeders need to pay attention to the breeding of plants in their ability to release cholines and, on the other hand, in resistance or sensitivity to cholines. So far, no-one has paid attention to this. It is interesting to note that in the development of ancient intensive agricultural, crops have gone through the creation of varieties and species of plants, which do not fatigue the soil. For example, rice, which has produced up to three harvests a year for hundreds or even thousands of years in one place, has no effect on the fatigue of the soil under irrigation and fertiliser use. In intensive culture on the South American continent, corn, potatoes, and tobacco can also be cultivated for centuries in one place, without causing deterioration of the soil. Grapes, common in the Mediterranean region, also do not significantly suffer from soil sickness.

However, in this, crops derived relatively recently from the wild state by humans, for example, sunflower, alfalfa, cotton, clover, mustard, often manifest pronounced soil sickness.

The same property is also shown in the relationship with plants that are growing together. For example, winter wheat, cultivated by man since time immemorial, is less aggressive in its relations with other plants than winter rye, which is similar in its ecology, but was much later introduced into cultivation. The same can be said about oats and barley.

Of course, the selection of plants, that do not fatigue the soil, has occurred earlier quite spontaneously and unconsciously. The characteristic of "fatigue-free" is unlikely to be correlated with other features, for which the selection is being conducted. Since the property of the soil fatigue impacts only the subsequent culture, it may be evident only in particularly severe developments. Selection could follow the line of the conservation of plants, in crop species tolerating crowding. On the other hand, since the property of soil sickness is related to the ability to counteract weeds, it acts in opposition to selection, that preserves plants that successfully withstand competition from weeds. In the future, apparently artificial selection is possible based on the knowledge of the causes of soil sickness. At the same time, depending on the purpose of cultivation, it is advisable to build up the properties of high or low allelopathic potential.

In agriculture, little attention is paid still to the quality of products, especially one that does not lend itself to be defined agrochemically or biochemically, and is assessed organoleptically – "bouquet" in wine, "flavour" of apples, baking properties of wheat and so on. These qualities are often defined by the relationships of allelopathic plants. Corn, and other grains that grow together with legumes, are rich in protein. Taste qualities, storability of fruits are higher in orchards with ground cover or in vineyards with grass, yet the harvest in these cases is worse. Potatoes take up the smell of hexachlorane, and reduce its taste in the presence of crude potassium salts. Obviously, with the knowledge of these matters, desired quality could arbitrarily be given to plants.

Allelopathy as the cycling of physiologically active substances in cenoses is among the number of powerful natural phenomena that determine the appearance of the vegetation, the composition of the microbial population of soil, and the structure, durability and productivity of phytocenoses.

Mastering the principles of allelopathy and their control by conscious human activity will be the step in the conquest of the elemental forces of nature. After the regulation of water and mineral nutrition of plants, the management of their ancestral properties, managing the interaction of plants in crops and natural cenoses is an important reservoir for further increasing the productivity of vegetation of the planet. Using various methods of farming that affect the conservation, transformation and degradation of cholines in the environment, by applying the allelopathic influences on cultivation, people can have even a more profound possession of natural plant resources.

APPENDIX (Applications)

METHODS OF STUDYING ALLELOPATHY

I. Methods for determination of cholines

The most important methodological issue during the allelopathic research is the determination of cholines - water-soluble and volatile physiologically active substances present in plant excretions. Sufficiently objective and accurate assessment could provide the physico-chemical and chemical methods. However, information on the chemical nature of cholines is too insufficient in its basis to allow a technique. Moreover, cholines almost always represent a very complex mixture of various physiologically interacting substances. Therefore, even when knowing the exact chemical composition of the complex solution, it is not always possible to predict in advance its biological effect.

In this regard, the method of biological assay is the principal and sometimes only method in allelopathy. Methods based on chemical and physical principles are still complementary, as they give an indication of the qualitative composition and quantity of substances that have certain biological effect.

A characteristic of the method of bioassay is that it has one major drawback, along with very positive attributes - simplicity, high efficiency, good reproducibility of results, and the ability to control more fully the conditions of experiments. It reproduces poorly the real relations taking place in nature, in the environment of natural phytocenoses. It is not infrequently found that an inhibitor in bioassay, in nature turns out to be a stimulant; substances that are very active, do not operate in nature, etc. This disadvantage is due to the fact that in a short laboratory experiment, it is difficult or impossible to take into account the influence of such factors:

(1) selective action of cholines in bioassay and as natural objects;

(2) alteration of cholines under the influence of specific microflora, the reaction of the medium, adsorption, temperature, light, humidity and other conditions that are different from the laboratory;

(3) interaction of the studied cholines with other allelopathic active substances, present in cenoses.

Therefore it is impossible to give absolute values to data from bioassays and short-term laboratory experiments, and transfer them without checking on the relations existing in cenoses. In addition, it is desirable not to be limited to any single bioassay, but to implement a system of biological tests, which are based on various physiological processes.

1. Bioassays on seed germination

One is to count the number of seeds sprouted per test solution compared to the germination of a control in water, when it is exactly 50%. As an experimental object we have chosen the seeds of radish of the variety "Red with white tip", because of all the tested crops they grow most quickly and uniformly. Usually it germinates 50% in 12 hours at a temperature of 26°C, and some time later - the other seeds. High germination energy is desirable, because the more active the growth processes taking place, the more sensitive the plant organism is to external influences (which is a general rule for any biological tests), and the shorter is the period of the experiment, the less the risk of microbial changes in the test solution. In addition, the shorter the term of the experiment, the higher the productivity and the less is the need for dishes and space.

Experiments have shown that even in the seeds of radish, but with very low vigour (old) of germination, germination to 50% took a few days, and they did not respond to quite harmful substances.

We use fresh, selected seeds with high germination energy, from the past year of reproduction. To improve the uniformity of seed, it should be sifted in sieves and used as a specific fraction (2.0-2.5 mm or 2.5-3.0 mm). The germination of radish is performed on filter paper in a Petri dish with a diameter of 9-10 cm. Thus in one dish, two hundred seeds are sown. So that each hundred is not mixed, a vertical fold in the filter paper is made along diameter of the dish, which divides the dish into two parts. Therefore, the filters were cut not round, but oval-shaped with a "reserve" for the fold.

Optimum wetting is achieved by the addition of 5 ml of water or test solution per dish. The amount of water 4-10 ml per dish does not significantly affect the results of germination.

Then radish is sown, usually in 6 hundreds (3 dishes) per treatment. Since sowing must be done rapidly, hundreds of seeds are counted in advance into tubes, from which they are then poured and uniformly distributed onto the paper. Experience has shown that one does not necessarily have to count the hundreds. If the seeds are sufficiently homogeneous, a measure can be made from a small segment of a glass tube with a diameter of 5-6 mm, which quantifies the portions of seeds. For each batch and fraction of seeds, one needs to prepare a new measure. This is done: one end of the tube is plugged, and a series of several specifically counted portions of the seed of the batch is then poured into the tube, for each a mark denotes the position of the top of the seed. Finding the average, one cuts off the end of the tube and uses it for measuring seeds. Each serving should contain 100 viable seeds. Therefore, if germinating seeds are below 100%, such as 95%, it is necessary that it contains a measured dose of just 105-106 seeds, or, in general:

$$A = \frac{100}{B} \times 100$$

Where: A - the number of seeds in measured portions;

B - germination in %

Radish was sown in Petri dishes, which were stored in the dark at 26°C for 11-12 hours, followed by the first counting of germination. If conditions do not allow maintenance of 26°C, and germination is conducted at room temperature, the period of the first count is postponed for a few hours. In any case, it produces a crop at the end of the effort, for the determination of germination the next morning. One should strive to determine emergence when it coincides with the moment when the seeds germinate in the control plates exactly at 50%. When the technique is sufficiently mastered, it is performed relatively easily, if the conditions of germination are more or less constant.

Germinated seed is considered as that where the radicle breaks through the seed coat: when counting the seeds, simply sort them with the tip of a pencil. Calculations must be performed quickly because radish germinates very energetically and slow work may affect the results. Tallying of germination in the experiment should not take more than 20 minutes, so one need to set the experiment to a maximum of 20 hundreds (3-4 treatments) for one count.

Then calculate the average germination of radish from the treatments and these are expressed in percentage germination at the appropriate to water that is taken as 100%. Conventionally, we call these figures "germination for K - 50%." This amount of control is taken in order to identify not only the inhibition, but also the stimulating effect. Moreover, when half of the seeds germinate, the activity of the germination process reaches a maximum velocity and hence maximal sensitivity to growth agents.

As a result, the calculation for each treatment is characterised by a single number, and, that is exactly seed germination vis-à-vis the control, at the moment when it reaches just 50% germination. Such a method is a mathematical expression of the inhibitor concentration in the aqueous solution that standardises the results and allows one to compare data among themselves and different experiments, but it is still inconvenient. The higher germination at K - 50%, the less so inhibitors in solution, at 100% there are none or the effect is masked, in transition to 100%, the solution is a stimulator of germination. At 0%, the amount of cholines is infinitely large. Such a varied interpretation of numbers hampers their comprehension, and further calculations. Since the chemical nature of cholines is diverse, and of uncertain activity in test solutions, we express them in conventional units, namely, in mg/l coumarin – a known inhibitor, that we have adopted as a standard.

The effect of different concentrations of coumarin is expressed as a typical unimodal curve with full inhibition at 1364 mg/l and sprouting stimulation at 5 mg/l. Trials with various preparations of coumarin gave quite matching curves. The value of the median deviation from the reference curve between individual measurements with four replications was 2.9%, which fully satisfies the requirements of precision in biological experiments. We have developed a scale for the conversion of germination of radish for K - 50% in conventional units of coumarin (RCU) as shown in Figure 47.

To determine volatile inhibitors, a samples of suitable material (3 g) on a glass plates 4 x 6 cm^2 is placed in Petri dishes, so that it does not touch the wet filter paper on the bottom. The sowing of radish, the sprouts produced, and counting are as described above.

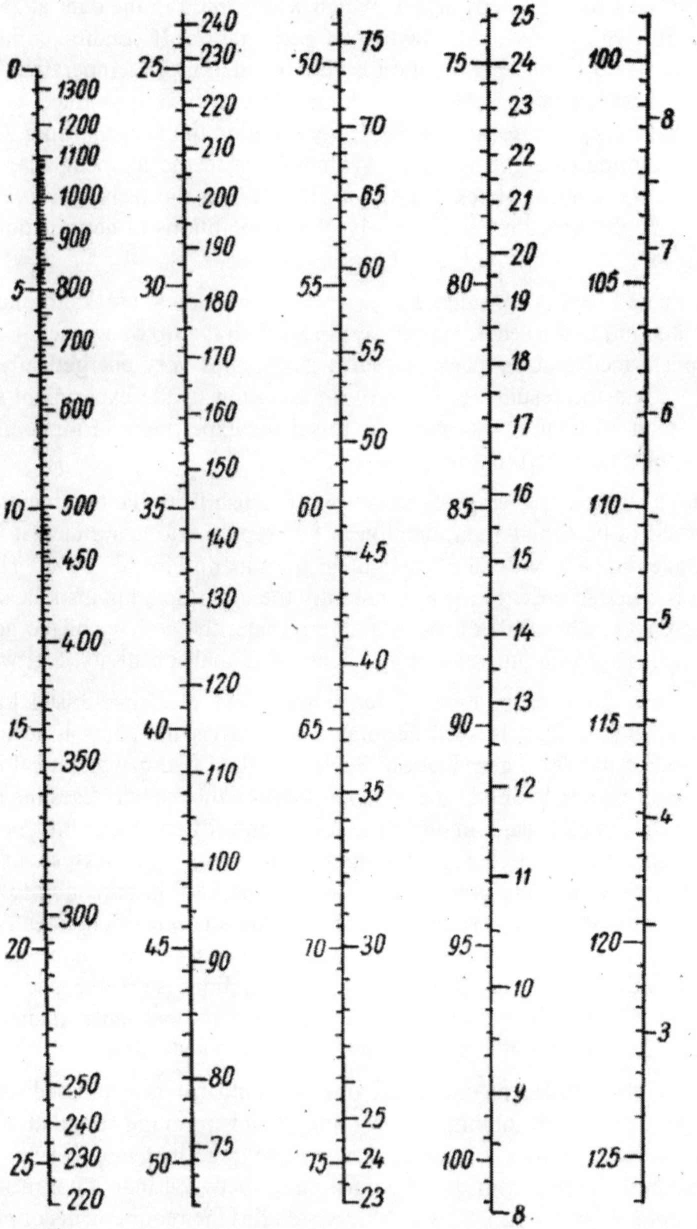

Figure 47. Scale for conversion of data from seed germination bioassay into arbitrary units of RCU. On the left side of the scale - the germination of radish for K – 50 %, on the right - RCU

A batch of material - the source of volatiles can also be put in a boat of foil attached to the side of the dish, or in a small cup or crucible, set in the center. In this case, one needs to cover the dishes with the same diameter lids, and ensure to seal the gap between them with adhesive tape.

A second, simplified, modification of bioassay has been developed by us for the small seeds, such as poppy, *Camelina*, shepherd's purse, *Erysimum*, wild radish, etc. A rectangular piece of paper 2.5 x 5 cm^2 connected in pairs with a crosspiece, placed at the bottom of the Petri dish, moisten 2 ml of water or solution and inoculated with poppy by using a small measure of 300-400 seeds. After 40-50 hours, visual assessment of seed germination and initial seedling development can be made. With this, the control always receives five points, and no germination is 0; the remaining options depend on the number of seeds and the total sprouts that form plantlets - from 0 to 5 and above, if the degree of germination is higher than the control. This method is less sensitive than the one described previously, but also quite accurate and suitable for biological display of chromatograms and mass rapid determination of cholines, if high accuracy is not needed. By comparing a large number of parallel determinations on radishes and poppy, we got a match rate of 0.78, which means the less sensitivity poppy (-ve sign) is in close agreement between the results in 78% of cases. For comparison of germination in points with germination percentage, one first needs to multiply by 20 to arrive at a number of the same order as percent germination.

It is recommended that several persons determine germination, and the result is then the average of their estimates.

Biological tests with radish and poppy seeds were very reliable and stable, giving standard results even when conducted in variable conditions. Through further counts of germination of radish for K - 50%, the impact of fluctuations in temperature and other conditions evened out. Re-test of the same extracts at different times of the year on the seeds of radish from different origins gave the same results. The same applies to the bioassays with poppy seeds.

This allows us to recommend the above bioassays not only for physiological, but also for experimental research on geo-botanical expeditions and in poorly equipped laboratories of reserves and hospitals. Bioassays on the seeds can also be used by farmers to determine cholines in fruits and seeds, in crop residues, local fertilisers, in the study of issues of soil sickness, etc.

These tests proved more reliable than the conventional bioassays with coleoptile segments, roots, and were much more productive.

A practical note: it is recommended that all labelling be done in pencil on the filter papers. This facilitates the subsequent washing of dishes, and eliminates confusion.

2. Bioassay on the roots of cress

Sorted cress seeds were plated on filter paper moistened with distilled water and placed in a dark incubator at 27°C. After a few days, when the seeds have germinated, select those whose roots have reached a length of 3 or 5 mm; 20 of them are placed in a Petri dish with the extract under study. After another day, measure the length of the roots and growth expressed as a percentage of the control growth of roots sprouted in water.

Measurement of root length is carried out conveniently by photography (A. Grodzinskii, D. Grodzinskii, 1964).

3. Bioassay with leaf discs

Bean, sunflower or pumpkin plants are grown in sand. From the primary bean leaves or green cotyledons of sunflower or pumpkin, discs with a diameter of 6-8 mm are cut with a cork borer. Leaves or cotyledons should be about the same age.

Discs are laid in a Petri dish on filter paper moistened with a 2% sucrose solution (control) or test solution containing 2% sucrose, and diffused light is allowed into the laboratory. Two days later the growth of control and experimental leaf discs is measured as the diameter, deaths are recorded, and leaf discs are macerated. If necessary, one can determine the decomposition of chlorophyll (fading), the strength of the chlorophyll-protein complex, the rate of photosynthesis with labelled C^{14}, etc.

4. Bioassay on segments of rhizomes of couch grass

Couch grass rhizomes are washed of soil, cleaned of scales and cut to a length of 1.5 cm, so that each segment contains dormant buds. Segments from 20 samples are laid on filter paper and moistened with water or test solutions. The assay is removed a few days later, when the control will begin to show growth of 50% of the buds. The percentage of sprouted buds, the number and length of new roots are determined.

We must bear in mind that at the end of July - August rhizomes of couch grass in parts of the southern region can be in the deep dormancy, and they cannot be used for bioassay.

5. Bioassay with rooting of cuttings from beans (A. Kh. Turtskoy)

Beans are grown in boxes or trays in the soil or sand at a temperature of 20-24°C and in sufficient light. When the seedlings reach a height of 11-13 cm which is usually at an age of ten days, the aerial parts are cut and the lower ends are immersed in the test solutions. For each, 50 cm^3 of solutions was poured into a beaker or wide flat-bottomed 200-250 cm^3 flask. After 4-6 hours, the cuttings are removed, rinsed and put into tap water. After 6-7 days, the intensity of root formation in the control and experimental plants is evaluated.

6. Bioassay with rooting of isolated cucumber cotyledons (per A.I. Beskhlebnyi)

Cucumber seeds are soaked overnight in water, the testa is removed, and about 1/3 of the seed along with the embryo is cut away. Remaining parts of the cotyledons are separated from each other and are laid on moistened filter paper in a Petri dish. Petri dishes are located in the laboratory in diffuse light. Two weeks later, the number of roots and their length are determined in the experimental and control plants.

7. Bioassay with the roots of white lupine

Selected dry seeds 9-11 mm in diameter are soaked for 24 hours in water at 20°C, in propagating sphagnum moss or digested oak sawdust moistened up to 300% by weight. Upon reaching a determined length of the roots (3 or 5 mm), seedlings are placed in a nutrient solution containing the test substance as well; 2-4 days later, root length is measured by expressing them as a percentage of growth of the roots in the control plants.

Similarly, one can use other large seeds – broad beans, peas, beans, pumpkin, sunflower.

8. Bioassay on roots (per Kholodnyi)

In maize seedlings, peas, sunflower and other plant roots are cut so that they have the same length - 2.5 or 3.0 cm (measurement made from the root tip to the plane of the slice). To obtain straight roots, the seedlings were grown in fine sifted sawdust, sphagnum peat, etc. Isolated roots are kept for several minutes in well water or weak Knop's solution, and then 10 pieces are arranged horizontally or obliquely on slides. Roots can be positioned vertically and easily by pinching them between two parallel slides. Root length is measured in two days (with longer time, the roots begin to die), and their status is checked in 7 days. During long-term experiments, the chamber should be ventilated from time to time.

9. Bioassays on volatile substances (per G. Molisch)

(a) Two to three-day old pea seedlings or seeds of vetch are grown in small pots, placed under a glass dome, wherein is located the source of gaseous products. A week later, the form of experimental and control plants (under a dome with a clean air) is compared. Indicators are: occurrence of delayed growth, etiolation, epinasty.

(b) Cut branches bearing leaves of Siberian peashrub and other shrubs or trees are placed in water and placed in a glass dome containing an atmosphere with the gaseous emissions. In the presence of active gas, the leaves fall as compared to the control.

10. Other biological methods

Many researchers, including us, have successfully used such indicators of the effects of volatile and water-soluble cholines on various plant functions:

(a) the velocity of the protoplasm;

(b) the permeability of protoplast;

(c) osmotic pressure and speed of onset of plasmolysis;

(d) the activity of various enzymes;

(e) the degree of opening of the stomata;

(f) the intensity of respiration and production of substances and photosynthesis..

11. Chemical methods for determining cholines

For qualitative and semi-quantitative characteristics of the group of the chemical composition of the group of plant excretions we use various methods of drop analysis set out in the relevant manuals (see; e.g., Feigl, 1962). In addition, droplets of exudates dried on chromatographic paper, the total content of amino acids, sugars, polyphenols, tannins, reducing agents, compounds that luminesce or absorb ultraviolet light, are determined. The amount of substances is evaluated in points.

II. Methods of extraction and capture of cholines

Allelopathy essentially involves water-soluble and volatile substances capable of moving in a cenosis. It is natural, therefore, that water is used as solvent at least for weak buffers, dilute solutions of alcohols, acids or alkalis. Lipophilic solvents should only be used in determining the chemical properties of cholines.

For extraction of physiologically active substances from various materials we use an infusion for 24 hours at room temperature in the following proportions of the sample in water: dry seeds and fruits 1:10; dry leaves stems, flowers, straw, stubble and other. 1:30 – 1:50; fresh intact leaves, fruits, roots, flowers and so on. 1:10; forest or steppe litter 1:4; soil 1:1.5.

To simulate the leaf secretions that occur during rain or dew, we apply the method, of so-called "removal"; 50 g of whole, intact leaves are placed in a wide-mouthed 500 cm^3 flask, 25 ml of water are added, and the moistened leaves are periodically gently shakened. After 30 min. water is poured into a graduated cylinder. Leaves typically absorb some water. Missing water is added to the leaves, which are rinsed and drained 1-2 times into the same cylinder.

Of great interest is the study of natural solutions: the leachate from lysimeters, water leaked through the hole in the bottom of the growth containers, nutrient solutions from hydroponics, water flowing from the leaves in the rain, dew, guttation, etc. To collect dew and guttation fluid, the use of well washed (boiled) gauze swabs is recommended. Drops collected from swabs are squeezed onto a Petri dish. One should wear thick rubber gloves.

For root exudates, we use the following methods:

1. In a water culture vessel is placed on the bottom 0.5-1 grams of activated carbon, pre-washed with 50% ethyl alcohol (ethanol or methanol) and dried. After 2-4 days, the activated carbon is removed, rinsed with water and eluted with 50% ethyl alcohol. The eluates were applied to a chromatogram or evaporated, and the residue was dissolved in water for testing in a bioassay.

2. Washed plants collected in the field are placed with their roots in Helrigel nutrient solution using aeration and frequent change of solutions every 3-4 days. Two weeks later, the root system is renewing itself: all the damaged

parts at this time die; these are removed when changing solutions, and new roots grow. The following occurs then with the plants from the water culture.

3. In water culture, the plants are transferred but roots are rinsed first in the tap water. The amount of water taken is 10-fold relative to the volume of the roots. Note the water level in the vessel and strive to maintain it all the time. After two days the water is drained and the solution is made up to that of the original root exudates concentration - 100%.

To study the volatile excretions from leaves, one variously collects the condensate of the transpiration fluid. In the simplest case it is sufficient to place the leaves, twigs or the entire aboveground parts in a fairly large flask. In hot sunny weather, the facility should be obscured with one or two layers of cheesecloth. The walls of the vessel will always be colder than the air inside it, so they will condense moisture, which then flows to the bottom. As shown in experiments by A.A. Musienko and us, condensation water has a high physiological activity.

More accurate results are obtained by freezing water vapor and cholines by pulling air through a fairly voluminous refrigerator at the temperature of dry ice or liquid nitrogen. S.S. Skvortsov, for example, studied substances deposited on the ice from lyophilised plant material.

To study volatile substances, it is most convenient to place the source of emissions into the chamber where the bioassay is conducted. The dosage concentration of volatiles is achieved by using defined samples. In a Petri dish with a volume of air 50 cm^3 we use these samples: crushed fresh leaves roots, stems, flowers and so forth (determination of volatile production) - 3 g soil; litter - 2 g + 1 cm^3 water; whole flowers, buds, leaves - 1-15 pieces, depending on their size; compost, organic fertilisers - 0.1-1.5 g.

Generally, the dosage concentration of cholines, because the chemical nature is unknown, presents greater difficulties. Apparently, one can gain some insight into the determination of total organic matter in a liquid or in the air by the method of oxidation using a solution of potassium dichromate in sulphuric acid, or potassium permanganate.

III. Methods of growing plants in experiments on allelopathy

By the use of special ways of positioning plants, allelopathic effects can be separated from other effects. It should be mentioned that these are techniques to be used in pot and field experiments.

1. Method of flow and circulating cultures

Plants were grown in aqueous culture in a hoop of plastic or stainless steel fixed above the nutrient solution. The rings resemble small sieves. They have an arbitrary diameter, height of 2-4 cm, and at the bottom on the outside, there is an annular groove for tying a nylon mesh. Large seeds (corn, peas, broad beans) are allowed to germinate beforehand in sand, and then are transplanted. In this hoop of nylon mesh, one moves apart the mesh with the tip of a pencil and the rootlets are placed in the openings, and the remaining space inside the ring is filled with gravel, glass or plastic beads. Medium-sized seeds (of wheat, rye, radish) after beginning to shoot are laid out on the mesh, and sprinkled with gravel. For small seeds (poppy, St. John's wort, etc.) one first pours onto the mesh a layer of gravel, then sand, in which seeds are sown. Rings are fixed over the water so that the mesh touches the solution, and germinating seeds are watered by the capillary rising of water, but it should not be flooded. This method of cultivation is very simple, easy to use, and provides homogeneous material. Accordingly, one can grow in water culture any plants, from desert succulents to trees.

As for nutrient solutions, any recommended in manuals on plant physiology can be used, especially where appropriate for crop species (A. Grodzinskii, D. Grodzinskii, 1964).

To study the allelopathic effect, one can either change the positions of plants after a certain period of their growth, or one can place vessels in sequence, on low steps to ensure continued weak flow of nutrient solution or by using small pumps to circulate the solution in a closed circle of several vessels. In conducting these tests, each treatment must be provided with a control that is suitable for comparison. For example, in the circulating system to study the interaction of plant A and plant B, at the same time as this, two such circulating systems must be delivered to plants A and B.

In the results of these experiments, there is imposed a significant character of disparity in the mineral nutrition of the plants, which may be impossible to avoid. For example, in studying the effect of root exudates of plant A on plant B, we give the plant B a solution, that to some extent has already been used as a solution without root exudates by A. Adjustment of the solution by adding the missing nitrogen, phosphorus, potassium, regulating the pH, does not achieve its goal completely, because there are still many other factors that are considered difficult.

The same principle is used in putting sand and soil culture in large funnels or vessels with holes in the bottom of the tray, and the irrigation water leaked from the 1st vessel is used for irrigation of the 2nd, etc. The series of vessels can have up to 10-15 elements, and the allelopathic effects in each successive vessel will increase. However, when watering, part of the water is retained in the vessel, and there is less liquid remaining for the irrigation of the next vessel. In this situation, one can make do, either by adjusting the volume of the filtrate until there is a constant flow of water flow, or by using more first vessels and gradually reducing their number towards the end of the series. For example, one can take for variable No.1 -20 vessels, No. 2 –16 vessels, No.3 –12 vessels, No.4 –8 vessels, No.5 -4 vessels

The foregoing about the impact of changes in mineral nutrition remains valid here.

2. Methods for the isolation of competition for water, light and nutrients

The essence of these methods is that the plants of the experimental variable are grown in the vessels with septa (or in two vessels of different sizes inserted into one another), so that their roots and aerial parts are in contact or separated. One can create either optimal conditions or conditions with insufficient moisture, mineral nutrition or light. In this way, for example, G. Grümmer was able to show that different types of false flax[1] oppress flax with secretions that are washed off the leaves by precipitation.

When the plants are isolated from each other, it not only stops the competition, but also some form of allelopathic effects, so one needs to use these methods very carefully.

3. The method of the biological chain

Used to study the migration of elements in the components of model phytocenoses. There are two ways.

At first, plants are planted in the sand or soil in tall flat-sided boxes with removable side walls, measuring 5 x 20 x 20 cm^3. After the plants have attained the required size, the side walls are removed, and the boxes are put together tightly in gangs, separated by two layers of filter paper. Plants from the first box are given the labeled compound (via a separate root, a non-root method or through photosynthesis with $C^{14}O_2$). After some time, its movement is determined in the other boxes. In watering produced from the last boxes, isotopes cannot be transported with the water flow.

With the second method, a gang is composed of wide tubes or vessels. Plants are planted at the "top" of junctions of vessels, so that half of the roots is in the same vessel, half - in the other. The labeled compound is introduced into the first vessel and is absorbed by the roots of the first strand of the plant, through the second strand it is released into the second vessel and is then absorbed by the following plants, etc. In all cases, it is necessary to take precautions to prevent accidental contamination and diffusion of substances from vessel to vessel.

Substances that move are fairly quickly diluted. Therefore, when the duration of the experiment is about a month, the gang cannot be more than 8-10 vessels (boxes).

4. The method of contacting and mixing the crop

This method entails plants that are planted in pure stands so that they abut each other. S.I. Chernobrivenko for field crops uses a small square plot located inside a solid background of sowing of the donor plant. The effect is determined by comparing the growth of the partners at the junction of plots and in their means where growth of one culture is on sites with different background cultures.

Very often it is used in mixtures of differing crops and density ratios of components, and compared with results from the same pure crop plants.

I must say that in these experiments, negative or positive interaction effects cannot always be attributed with certainty to allelopathy, as the competition for

---

[1] Grodzinskii gives the name of this plant as *рыжик*, which is the name for the mushroom known as saffron milky cap (*Lactarius deliciosus*). This is undoubtedly an error, as the name intended was the similar *рыжей*, which is the name for false flax (*Camelina* spp.), with which Grümmer worked.

water, nutrients cannot be eliminated. V.N. Sukachev and V.G. Karpov used digging of the roots at the boundaries of the plots. This enables one to get rid of root competition and, as shown by numerous experiments, gives very clear results. However, in this case there are dying severed roots in the soil plots, which are not always taken into account by researchers.

On occasion, when the study of the allelopathic effect of components is of unequal power, such as trees and grasses, it can be undertaken in two ways, the nature of which is clear from Figure 48.

Each plot is the experimental and the control. In the presence of allelopathic effect, the test plants that are located near the donor must be suppressed; according to the distance, their growth will improve. Finally, at a certain distance, where the concentration of choline reaches the optimal value, it should be observed that there is stimulation of growth in the culture. The latter is the best proof of allelopathy, because competition may rarely cause growth stimulation.

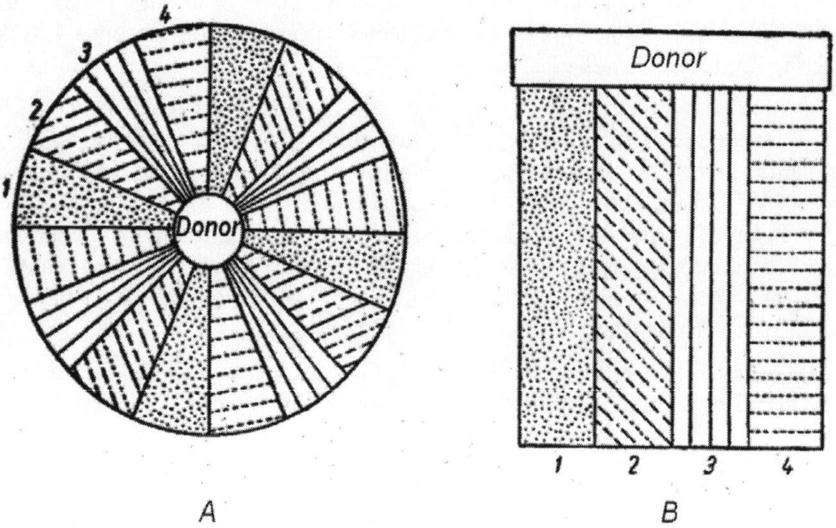

Figure 48. Layout scheme of experimental crop plants relative to the donor: A - Donor in the centre: B – Donor at on one side. 1,2,3,4 - plots of different crop plants.

REFERENCES

Aamisepp, A. and Osvald, H. 1962. Influence of higher plants upon each other – allelopathy. Some new results of research into allelopathy, etc. *Nova Acta Regiae Societatis Scientiarum Upsaliensis,* Ser. 4, 18(2): 1-19.

Ado A. D. 1961. [Allergy]. *Priroda,* 1961 (1): 61-65.

Aksent'yev B. N. 1927. [The influence of seed extracts on seed germination] *Zhurnal Russkago Botanicheskago Obshchestva pri Akademii Nauk* 12(3): 291-314.

Aleksandrova V.D. 1963. [Experiments in analysing the phenomena of self-regulation in phytocenosis in terms of some of the ideas of cybernetics]. In: *Primeneniye Matematicheskikh Metodov v Biologii.,* Vypusk 2. Izd-vo Leningradskogo Universiteta, Leningrad. Pp. 37-46.

Alekhin, B.B. 1939. [Some thoughts on the life forms of "tumbleweed"]. In: Prezidentu Akademii Nauk SSSR Akademiku Vladimiru Leont'yevichu Komarovu k Semidesiyatiletiiyu so Dnya Rozhdeniya i Sorokapyatiletiiyu Nauchnoi Deitel'nosti. Akademiya Nauk SSSR, Leningrad. Pp. 97-107.

Andal, R., Bhuvaneswari, K., and Subba-Rao, N.S. 1956. Root exudates of paddy. *Nature* 178: 1063.

Anderson, G. 1957. Nucleic acid derivatives in soils. *Nature* 180: 287-288.

Arens, K. 1934. Die kuticuläre Exkretion de Laubblattes. *Jahrbücher für wissenschaftliche Botanik* 80(3): 248-300.

Arens, K. 1936a. Physiologisch-polarisierter Massenaustausch und Photosynthese bei submersen Wasserpflanzen. II. Die $Ca(HCO_3)_2$-Assimilation, *Jahrbücher für wissenschaftliche Botanik.*83(4): 513-560.

Arens, K. 1936b. Kann man in allen Fallen aus den Ergebnissen der Aschenanalyse Schlüsse auf Lebensfunktionen der Pflanzen ziehen? *Landwirtschaftliche Jahrbücher* 82(3): 453-463.

Bahme, R. B. 1949. Nicotinic acid as a growth factor for certain orchid embryos. *Science* 109: 522-523.

Bergh, J.P. van den and Elberse, W. T. 1962. The competition between *Lolium perenne* L. and *Anthoxanthum odoratum* L. at two levels of phosphate and potash. *Journal of Ecology* 50(1). 87-95.

Beskaravaynyy, M.M. 1955. [Fusion of roots of some trees near the city of Kamyshin]. *Agrobiologiya* 1955(3): 9-12.

Bhan, K.C., Wallace, A. and Lunt, O.R. 1959. Some mineral losses from leaves by leaching. *Proceedings of the American Society of Horticultural Science.* 73: 289-293.

Bilay, V.I. 1952. [Effect of different species of *Fusarium* on cereals] [Ukrainian]. *Mikrobiolohichnyy Zhurnal* 14(4).

Bilay V.I. 1955. [Symbiotrophic properties of *Fusarium* species and other soil fungi]. In: *Trudy Konferentsii po Mikotrofii Rastenii.* Ed. Imshenetskij, A.A. Izd-vo Akademii Nauk USSR, Moscow (This work was translated into English in 1967, and appeared in *Mycotrophy in Plants*, Program for Scientific Translations, Jerusalem, Pp. 129-141).

Blagoveshchenskiy A. V. 1951. [On substances that delay germination]. *Byulletin Glavnogo Botanicheskogo Sada Akademii Nauk SSSR* 1951 (9): 54-58.

Blagoveshchenskiy A. V. 1953. [Biochemistry of the difficult germination of seeds], *Trudy Glavnogo Botanicheskogo. Sada, Akademii Nauk SSSR*, 1953(3): 3-57.

Blagoveshchenskiy A. V. 1962. *Biogennyye Stimulya i Urozhay.*[=Biogenic stimulators and yield]. Izd-vo Znaniye, Moscow.

Blaim, K. 1955. Studia nad biochemia kiekujacego ziarna pszenic ozimych i jarych. *Roczniki Nauk Rolniczych*, ser. A, 72(1): 9-37.

Blaim, K. 1960. The effect of coumarin on uptake of water by seeds. *Journal of Experimental Botany* 11(3): 377-380.

Blanchard, F. A. and Diller, V. M. 1951. Uptake of aureomycin through the roots of Phaseolus lunatus. *American Journal of Botany* 38: 111-112.

Bode, H. R. 1940. Über die Blattausscheidungen des Wermuts und ihre Wirkung auf andere Pflanzen. *Planta* 30(5): 567-589.

Bode, H. R. 1958. Beiträge zur Kenntnis allelopathischer Erscheinungen bei einigen Juglandaceen. *Planta* 51(4): 440-480.

Bonner, J. and Galston, A. W. 1944. Toxic substances from the culture media of guayule which may inhibit growth. *Botanical Gazette* 106: 185-198.

Bonner, J. 1950. The role of toxic substances in the interactions of higher plants. *Botanical Review* 16(1): 51-65.

Börner, H. 1955. Die ausscheidungen organischer Verbindungen aus den Samen von Roggen (*Secale cereale* L.), Weizen (*Triticum aestivum* L.) und Gerste (*Hordeum vulgare* L.) während des Quellung. *Naturwissenschaften* 42(2): 583-584.

Börner, H. 1956a. Der papierchromatographische Nachweis von Ferulsäure in wässrigen Extrakten von Getreidestroh und Getreiderückständen. *Naturwissenschaften* 43(6): 129-130.

Börner, H. 1956b. Die Abgabe organischer Verbindung aus den Karyopsen, Wurzeln und Ernterückständen von Roggen (*Secale cereale* L.), Weizen (*Triticum aestivum* L.) und Gerste (*Hordeum vulgare* L.) und ihre Bedeutung bei der gegensetige Beeinflusung der höheren Pflanzen. *Beiträge zur Biologie der Pflanzen* 33(1): 33-83.

Börner, H. 1958a. Experimentelle Untersuchungen zum Problem der gegenseitigen Beeinflussung von Kulturpflanzen und Unkräuter. *Biologisches Zentralblatt* 77: 310-328.

Börner, H. 1958b Nachweis phenolischer Verbindung in Leinsamen und ihre Abgabe während der Quellung. *Flora* 145: 479-496.

Börner, H. 1958c. Untersuchungen über den Abbau von Phlorizin im Boden in Beitrag zum Problem der Bodenmüdigkeit bei Obstgehölzen. *Naturwissenschaften* 47: 138-139.

Börner, H. 1960a. Liberation of organic substances from higher plants and their role in the soil sickness problem. *Botanical Review* 26(3): 393-424.

Börner, H. 1960b. Über die Bedeutung gegenseitiger Beeinflussung von Pflanzen in landwirtschaftlichen und forstlichen Kulturen. *Angewandte Botanik* 34(3/4): 192-211.

Börner, H. and Rademacher, B. 1957a. Experimentelle Untersuchungen zum Problem der Selbverträglichkeit der höheren Pflanzen. *Naturwissenschaften* 44: 498-499.

Börner, H. and Rademacher, B. 1957b. Untersuchungen zum Problem der echten Selbstunverträglichkeit des Leins (*Linum usitatissimum* L.). *Zeitschrift für Pflanzenernährung, Dungung, Bodenkunde* 76(2): 123-132.

Bowen, G.D. 1961. The toxicity of legume seed diffusates toward rhizobia and other bacteria. *Plant and Soil* 15: 155-165.

Braun-Blanquet, J. 1951. *Pflanzensoziologie. Grundzüge der Vegetationkunde.* Springer Verlag, Vienna.

Brehmer, D.E. and McCalla, T.M. 1963. The inhibition of seedling growth by crop residues in soil innoculated with *Penicillium urticae. Plant and Soil* 18: 199-206.

Brian, P.W., Wright, J.M, Stubbs, J., and Way, A.M. 1951. Uptake of antibiotic metabolites of soil micro-organisms by plants. *Nature* 167: 347-349.

Bublitz, W. 1954. Die Bedeutung von Hemmstoffen in der Forstwirtschaft. II. Mitteilung: Untersuchungen über den Nachweis antibiotisch wirksamer Fichtenrohhumussubstanzen und ihr Einfluss auf die Entwicklung von Bodenbakterien. *Madaus Jahresbericht* 7: 92-106.

Bykov B.A. 1961. [An interesting fact of environmental influence (allelopathy)], *Botanicheskie Zhurnal* 46(1): 108-112.

Candolle, A.P. de 1832. *Physiologie Végétale.* Béchet Jeune, Paris.

Cannon, J.R., Corbett, N.H., Haydock, Tracey, J.G. and Webb, L.J. 1962. An investigation of the effect of the dehydroangustione present in the leaf litter of *Backhousia angustifolia* on the germination of *Araucaria cunninghamii* – An experimental approach to a problem in rain-forest ecology. *Australian Journal of Botany* 10(2): 119-128.

Carré, M. 1959. Nature et rôle de quelques substances à fonction acide excrétées par les radicelles des végétaux supérieurs. *Bulletin de la Société Botanique de France* 106(1/2): 13-16.

Cato, Varro, Columella and Pliny. 1937. *O Sel'skom Khozyaystve* [=On Agriculture]. Sel'khozgiz, Moscow.

Cézard, N. 1960. Excrétions racinaires. *Bulletin de la Société des Sciences de Nancy* 19(4): 234-248.

Cézard, R. 1961. L'excrétion radicellaire chez les végétaux supérieurs. *Bulletin de l'École Nationale Supérieure de l'Agronomique de Nancy* 3(2): 114-166.

Chailakhyan, M.Kh. 1961. *Gibberelliny Rastenii* [=Gibberellins of Plants]. Izd-vo Akadamii Nauk SSSR, Moscow.

Chailakhyan, M.Kh. 1962. [Metabolites of microbes as stimulators of growth and development in higher plants]. In: *Primenenie Antibiotikov v Rastenievodstve*, Izdanie Akademii Nauk Armyanskoi SSR, Erevan. Pp. 113-121.

Chalvignac, M.A. 1958. Effet rhizosphère comparé du lin en culture hydroponique et en terre. *Annales de l'Institut Pasteur* 95: 474-479.

Chalvignac, M.A. 1962. Étude biochimique des sécrétions radicellaires du lin en culture hydroponique. *Annales de l'Institut. Pasteur* 102(9): 767-770.

Chasovennaya, A.A. 1954. [The effect of volatile organic excretions of plants on the germination of seeds, growth and development of some species of herbaceous plants] *Vestnik Leningradskogo Gosudarstvennogo Universiteta* 1954(1): 25-31.

Chasovennaya, A.A. 1961 [On the mechanism of the chemical interaction of plants]. *Vestnik Leningradskogo Gosudarstvennogo Universiteta, Seriya Biologiya* 1961(3): 54-66.

Chernobrivenko, S.I. 1956. *Biologicheskaya Rol' Rastitel'nykh Vydelenij i Mezhvidovye Vzaimootnosheniya v Smeshannykh.* [=Biological Role of Plant Secretions and Interspecific Relations in Mixed Crops]. Sovetskaya Nauka, Moscow.

Constabel, F. 1963. Quantitative Untersuchungen über die extracellulaire Hydrolyse von Kohlenhydraten durch *Juniperus communis* – Gewebekulturen. *Planta* 59(3): 330-337.

Curtis, L.C. 1944a. The influence of guttation fluid on pesticides. *Phytopathology* 34: 196-205.

Curtis, L.C. 1944b. The exudation of glutamine from lawn grass. *Plant Physiology* 19(1): 1-5.

Curtis, R.W. 1961. Studies on response of bean seedlings and corn roots to malformin. *Plant Physiology* 36(1): 37-43.

Dadd, C. C., Fowden, L., and Pearsall, W.H. 1953. An investigation of free amino-acids in organic soils types using paper partition chromatography. *Journal of Soil Science* 4(1): 69-71.

Deleuil, G. 1950. Mise en evidence de substances toxiques pour les thérophytes dans les associations du Rosmarino-Ericion. *Comptes Rendus de l'Académie des Sciences* 230: 2476-2477.

Deuel, H. 1959. Polysaccharide im Boden. *Angewandte Chemie* 71: 196.

Dommergues, Y. 1963. Les cycles biogéochimiques des éléments minéraux dans les formations tropicales. *Bois et Forets des Tropiques* 83: 9-25.

Drabkin, B.S. 1954. [On the mechanism of action of phytoncides from wild cherry]. *Biokhimiya* 19(4): 385-390.

Drobotko, V.G., Aizenman, B.E., Shvaiger, M.O., Zelepukha, S.I, and Mandrik, T.P. 1958. *Antimikrobnye Veshchesva Vysshykh Rastenii* [=Antimicrobial substances in higher plants]. Izd-vo Akademii Nauk USSR, Kiev.

Dubovik, Ya.F. 1950. [The role of biological absorption of mineral salts in the soil in the nutrition of plants]. *Doklady Akademii Nauk SSSR* 73(4).

Duquenois, P. 1955. Les antibiotiques des plantes supérieures. *Bulletin Société Botanique de France* 102(7/8): 377-405.

Duquenois, P. and Schaelderlé, D. 1958. Revue complémentaire sur les antibiotiques des plantes supérieures. *Bulletin Société Botanique de France* 105(9/10): 526-560.

Dykyj-Sajfertova, D and Dykyj, J. 1956. The influence of growth stimulators on the germination of crop seeds at different pH. *Preslia* 28(1): 1-22.

Eberhardt, F. 1954. Ausscheidung einer organisch Verbindung aus den Wurzeln des Hafers (*Avena sativa* L.). *Naturwissenschaften* 41 (11): 259.

Eberhardt, F. 1955. Über fluoreszierende Verbindungen in der Wurzel des Hafers. Ein Beitrag zum Problem der Wurzelausscheidungen. *Zeitschrift für Botanik* 43 405-422.

Ellenberg, H. 1953. Physiologisches und ökologisches Verhalten derselben Pflanzenarten. *Berichte der Deutschen botanischen Gesellschaft* 65: 351-362.

Engel, H. 1939. Das Verhalten der Blätter bei Benetzung mit Wasser *Jahrbücher für wissenschaftliche Botanik* 88: 816-861.

Evenari, M. 1949. Germination inhibitors. *Botanical Review* 15(3): 153-194.

Fedorovskij, D.V.O. 1958. [On the excretion of labeled phosphorus and calcium from roots in the soil in root fed plants]. *Pochvovedenie* 1958(3): 17-23.

Filatov, V.P. 1951. [Tissue treatment (the study of biogenic stimulators)]. *Priroda* 1951 (11): 39-43.

Filippovich, T.N. 1962. [On the influence of root exudates of brome grass on the supply of nutrients to plants of couch grass].. In: *Tezisy Dokladov Nauchnoi Konferentsii po Voprosam Eksperimentel'noi Geobotanike*. Izd-vo Kazanskogo Universiteta, Kazan'.

Filippovich, T.N. 1963a. [The influence of root secretions on the uptake of P^{32} and S^{35} in young plants of oats and brome grass] [Ukrainian] In: *Fiziologichnoi-Biokhimichnoi Osnovi Pidvishcheniya Produktivnosti Roslin*. Derzhsil'gospvidav URSR, Kiev.

Filippovich, T.N. 1963b. [The dependence of the process of absorption of nutrients on the effects of the plant root secretions] [Ukrainian]. In: *Pytannya, Fiziolohiyi, Tsytoembriolohiyi i Flory Ukrainy* [=Nutrition, physiology, cytoembryology and the flora of Ukraine]. Vid-vo Akademiya Nauk, URSR, Kiev.

Flaig, W. and Breyhan, H. 1956. Über das Vorkommen von Indolverbindungen in Schwarzerde-Huminsäuren. *Zeitschrift für Pflanzenernährung, Dungung, Bodenkunde* 75(2): 135-143.

Flaig, W. and Kaul. R. 1956. Untersuchungen über die Beeinflussung der Phosphat aufnahme von Haferpkeimflanzen durch Phosphatasegifte und Phosphataseaktivatoren. *Zeitschrift für Pflanzenernährung, Dungung, Bodenkunde* 72(2): 97-103.

Flaig, W. and Saalbach, E. 1959. Über den Einfluss von im Boden gerottetem Stroh auf das Wachstum und die Nährstoffaufnahme von Roggenkeimpflanzen. *Zeitschrift für Pflanzenernährung, Dungung, Bodenkunde* 87(3): 229-235.

Fraenkel, G.S. 1959. The raison d'être of secondary plant substances. *Science* 129(3361): 1466-1470.

Frank, H. 1954. Über den Stickstoffverlust bei alternden Pflanzen. *Planta* 44(4): 319-340.

Fransson, P. 1960. An active state of auxin in wheat roots. *Physiologia Plantarum* 13(2): 398-428.

Frenzel, B. 1957. Zur Abgabe von Aminosäuren und Amiden an das Nährmedium durch die Wurzeln von *Helianthus annuus* L. *Planta* 49(2): 210-234.

Frenzel, B. 1961. Modellversuche zur Klärung einiger Probleme der Abgabe von Aminoverbindungen durch die Wurzeln höherer Pflanzen, im Boden. *Planta* 57(4): 444-454.

Frey-Wyssling, A. 1949. *Stoffwechsel der Pflanzen*. Büchergilde Gutenberg, Zurich.

Fries, N. and Forsman, B. Qualitative determination of certain nucleic acid derivatives in pea root exudate. *Physiologia Plantarum* 4(3): 410-442.

Gander, J. E. 1960. Growth response of *Triticum* coleoptiles to non-dialysable milk fractions. *Journal of Experimental Botany* 11(3): 351-355.

Garb, S. 1961. Differential growth-inhibitors produced by plants. *Botanical Review* 27: 422-443.

Gäumann, E. and Jaag, O. 1946. Über das Problem der Welkekrankheiten bei Pflanzen. *Experientia* 2(6): 215-220.

Gautheret, R. J. 1959. La Culture des Tissus Végétaux. Techniques et Réalisations. Masson et Cie, Paris.

Geller, I.A. and Tabentskij, D.A. 1957. [Root exudates and plant nutrition]. *Doklady Akademii Nauk SSSR* 115(2): 21-25.

Gellerman, Ya. M 1950. [The effect of the products of decomposition of organic substances on the root nutrition of plants]. *Doklady Akademii Nauk SSSR* 73(3): 593-596.

Gel'man, A.A.1961. [The effect of phytoncides from some wild pasture plants on seed germination and initial growth of seedlings of plants sown in pasture grass mixtures]. *Trudy Vologodskogo Molochnogo Instituta* 44: 33-37.

Golomedova, T.I. 1952. [Mutual toxicity of plants due to their aqueous extracts] *Agrobiologiya* 1952(2): 132-134.

Golomedova, T.I. 1954. [More on the mutual toxicity of plants due to their aqueous extracts] *Agrobiologiya* 1954(2): 125-128.

Golubinskii, I.N. 1946. [On the mutual influence of different types of pollen grains when sharing their germination in artificial media]. *Doklady Akademii Nauk SSSR* 53(1): 73-76.

Golubinskii, I.N. 1951. [Influence of contaminating pollen of surrounding plants on the germination of pollen]. *Doklady Akademii Nauk SSSR* 76(4): 591-593.

Gortinskii, G.B. 1963. [Allelopathy in biocenology (about the biocenological approach to the study of allelopathy)]. *Byulleten' Moskovskogo Obshchestva Isputatelei Priroda, Otdel Biologicheskii* 68(6): 103-110.

Gray, R. A. 1955. Inhibition of root growth by streptomycin and reversal of the inhibition by manganese. *American Journal of Botany* 42(3): 327-331.

Greenhill, A.W. and Chibnall, A.C. 1934. The exudation of glutamine from perennial rye-grass. *Biochemical Journal* 28: 1422-1427.

Grineva, G.M. 1961. [On the secretion of substances by plants under brief anaerobiosis]. *Fiziologiiya Rastenii* 8(6): 686-691.

Grodzinskii, A.M. and Grodzinskii, D.M. 1964. *Kratkiy Spravochnik po Fiziologii Rastenii* [=A short handbook of plant physiology]. Naukova Dumka, Kiev.

Grodzinskii, A.M., Kuznetsova, G.O. and Musatenko, L.I. 1960. [Inhibitors of germination from fruits of Tatarian bread *Crambe tataria* Sebeök.] [in Ukrainian]. *Ukrainskii Botanicheskii Zhurnal* 17(1): 29-39.

Grodzinskii, A.M. and Osichnyuk, V.V. 1962. [Guttation of plants under natural conditions in Ukraine] [in Ukrainian]. *Ukrainskii Botanicheskii Zhurnal* 19(4).

Grodzinskii, A.M., Mokhova, N.I., Pilipenko-Yurchak, L.D., and Filipovich, T.M. 1962. [On inhibitory substances in stubble field crops and weeds. 1. Effect of soluble inhibitors on seed germination and plant growth]. *Ukrainskii Botanicheskii Zhurnal* 19(6): 30-38.

Grümmer, G. 1955. Die gegenseitige Beeinflussung höherer Pflanzen – Allelopathie. Gustav Fischer, Jena.

Grümmer, G. 1957a. *Vzaimnoe Vliyaniya Vysshikh Rastenii – Allelopatiya.* [The Mutual Interaction of Higher Plants – Allelopathy]. Izd-vo Inostrannoi Literatury, Moscow. (N.B. this is the Russian translation of Grümmer (1955)).

Grümmer, G. 1957b. Neuere Erkenntnisse über die gegenseitige Beeinflussung höherer Pflanzen. *Wissenschaftliche Zeitschrift* der Ernst-Moritz-Arndt-*Universität Greifswald.* Mathematisch-naturwissenschaftliche Reihe . 6: 245-250.

Grümmer, G. 1958. Die Beeinflussung des Leinertrages durch *Camelina* – Arten. *Flora* 146(1/2): 158-177.

Grümmer, G. 1961. The role of toxic substances in the interrelationships between higher plants. In: *Mechanisms in Biological Competition*, Ed. J. L. Harper et al. *Symposium of the Society for Experimental Biology* 15: 219-228.

Gulyaev, E.I. 1963. *Smeshannyye posevy kukuruzy s odnoletnimi bobovymi kul'turami v usloviyakh yuzhnoy stepi Ukrainy.* [Mixtures of corn crops with annual legumes in the conditions of southern steppe of Ukraine]. Thesis, Odesskiy Sel'skokhozyaystvennyy Institut, 28 pp.

Guminski, S., Czerwinski, W., Unger, E. and Bacowa, A, 1955. Studies on the respiration of the root. I.. *Acta Societatis Botanicorum Polaniae.* 24(4): 723-731.

Guyot, L. and Becker, Y. 1951. Sur les toxines racinaires des sols incultes. *Comptes Rendus de l'Académie des Sciences* 232: 105-107.

Guyot, L. Becker, Y., Massenot, M. and Montegut, J. 1951. Sur la présence d'excrétats radiculaires toxiques dans la pelouse herbeuse à *Brachypodium pinnatum* du nord de la France. *Comptes Rendus de l'Académie des Sciences* 231: 165-167.

Guyot, L., and Massenot, M. 1950. Sur la persistance prolongée de semences dormantes dans le sol de la pelouse herbeuse à *Brachypodium pinnatum* du nord de la France. *Comptes Rendus de l'Académie des Sciences* 230: 1894-1896.

Gvamichava, N.E. 1963. *Vitaminy v Nekotorykh Pochvakh Gruzii ikh Dinamika i Proishozhdenie* (=Vitamins in some soils of Georgia and their performance and origin). Nauka, Tbilisi.

Hamence, J.H. 1945. The occurrence of auxins in organic manures. *Journal of the Society of Chemical Industry* 64: 147-148.

Harley, J.L. 1952. Associations between microorganisms and higher plants (mycorrhiza). *Annual Review of Microbiology* 6: 367-386.

Heller, R. 1955. De quelques composés du soufre assimiables par les tissues végétaux cultivée *in vitro*. *Comptes Rendus de l'Académie des Sciences* 241 (2): 234-236.

Hohn, K. and Elfert, A.. 1958. Guttation, Atmung, Spross- und Wurzelwachstum unter Einwirkung ätherischer Öldämpfe auf die Wurzeln von Hafer. *Beiträge zur Biologie der Pflanzen* 34(2): 247-266.

Hull, A.C. Jr. 1963. Competition and water requirements of cheatgrass and wheatgrasses in the greenhouse. *Journal of Range Management* 16(4): 199-204.

Hutchinson, H.B. and Miller, N.H.J. 1911. The direct assimilation of inorganic and organic forms of nitrogen by higher plants. *Centralblatt für Bakteriologie, Parasitenkunde und Infektionskrankheiten. II.* 30: 513-547.

Isip, I.N. 1938. [The quiet and intensity of plant development]. *Selektsiya i Semenovodstvo* 7: 35-38.

Isip, I.N. 1939. [Influence of protective substances of the plant organism on seed germination]. *Sovetskaya Botanika* 3: 3-8.

Ivanov, V.P. 1962. [Mutual influence of corn and broad beans through the root system when sown together]. *Fiziologiya Rastenii* 9(2): 179-188.

Ivanov, V.P. 1963. [Is self-poisoning possible through plant root exudates?] *Doklady Akademii Nauk SSSR* 149(6): 76-79.

Ivanov, V.P., Yakobson, G.A. and Fomenko, B.S. 1963. [The mutual influence of plants of maize and broad beans through their aerial parts]. *Fiziologiya Rastenii* 10(4): 19-23.

Kakhidze, N.T. and Medvedeva, G.A. 1956. [The study of the excretion of vitamins by the flowering organs by means indicator cultures]. *Fiziologiya Rastenii* 3(5): 435.

Kandler, O. 1951. Papierchromatographischer Nachweis der Aminosäure-ausscheidungen *in vitro* kultivierter Maiswurzeln. *Zeitschrift für Naturforschung B* 6: 437-445.

Katznelson, H., Rouatt, J.W., and Payne, T.M.B. 1955. The liberation of amino acids and reducing compounds by plant roots. *Plant and Soil* 7(1): 35-48.

Kaul, R. 1962. Über die Ursachen der Wuchsförderung von *Stellaria media* durch Rottesubstanz im Boden. *Beiträge zur Biologie der Pflanzen* 37(2): 165-181.

Kefeli, V.I. and Turetskaya, R.Kh. 1963. [On some properties of vegetative inhibitors of plants]. Fiziologiya Rastenii 10(1)[1]

Kharchenko, S.M. 1961. [Antibiotic properties of species of the section *Monoverticillata* of the genus *Penicillium* removed from the rhizosphere of crop plants in Ukraine. III. Effect of species from the section *Monoverticillata*, genus *Penicillium*, on the growth of cabbage seeds and on certain bacterial plant diseases].

Mikrobiolohichnyy Zhurnal 23(1): 46-50.

Kholodnyj, N.G. 1957. *Izbrannye Sochineniya T.3* [=Collected Works. V. 3]. Izd-vo Akademiya Nauk SSSR, Moscow.

Khristeva, L.A. 1948. [The nature of the impact of humic acids on plants]. *Doklady Vsesoyuznaya Akademiya Sel'skokhozyaystvennykh Nauk imeni Lenina* (7): 131-136.

Khristeva, L.A. 1951. [The role of humic acids in plant nutrition and humic fertilisers]. *Trudy Pochvennogo Instituta im. V.V. Dokuchaeva* 38:108-184.

Khristeva, L.A. 1953. [On the participation of humic acids and other organic substances in the nutrition of higher plants]. *Pochvovedenie* 48(10): 46-59.

---

[1] This reference seems to be in error, and the appropriate reference is likely: Turetskaya, R.Kh. and Kefeli, K. I. 1963.[On some properties of natural inhibitors on the growth of plants]. *Fiziologiya Rastenii* 10 (1): 98.

Khristeva, L.A.. 1955. [The participation of humic acids and other organic substances in the nutrition of higher plants and agronomic value of this kind of nutrition]. *Izvestiya Akademii Nauk SSSR, Seriya biologicheskoi* (4): 58-83.

Khristeva, L.A., Pivovarov, L.R., Pshenichnyi, A.E. and Yarchuk, I.N. 1954. [Relationship between mineral and organic plant nutrition and the use of humic acids as fertiliser]. *Pochvovedenie* (12): 1-10.

Kingsley, A.F., Clagett, C.O. , Klosterman, J., and Stoa, T.E. 1957. Loss of minerals from mature wheat and flax by simulated rain. *Agronomy Journal* 49(1): 37-39.

Kirillova, T.V. 1964. [The secretion of water-soluble substances by plant tissues]. *Uspekhi Sovremennoi Biologii* 57(3): 396-401.

Kloke, A. 1963. Die Humusstoffe des Bodens als Wachstumsfaktoren. P. Parey, Berlin.

Knapp, R. 1954. Experimentelle Soziologie der höheren Pflanzen. E. Ulmer, Stuttgart.

Knapp, R. 1960a. Die gegenseitige Beeinflussung von Pflanzen in natürlichen Vergesellschaftung. *Angewandte Botanik* 34(3/4): 179-191.

Knapp, R. 1960b. Wirkung von mit Gibberellin-säure behandelten Pflanzen auf neben ihnen wachsende Individuen. *Naturwissenschaften* 47(12): 285-286.

Kolesnikov, A.V. 1962. Kornevaya sistema plodovykh i yagodnykh rasteniy i metody yeye izucheniya [=Root systems of fruit and berry plants and methods for their study]. Sel'skhozgiz, Moscow.

Kolesnichenko, M.V. 1960a. [The influence of the excretions of white birch on photosynthesis in oak]. *Doklady Akademii Nauk SSSR* 132(1): 220.

Kolesnichenko, M.V. 1960b. [On the question of the biochemical mutual relations of tree species]. *Lesnoi Zhurnal* 5: 37.

Kolesnichenko, M.V. 1961a. [On the chemical interrelation of species in forestation]. *Lesnoe Khoyaistvo* 11: 36-49.

Kolesnichenko, M.V. 1961b. [On intraspecific relations in English oak]. *Doklady Akademii Nauk SSSR* (3): 740.

Kolesnichenko, M.V. 1962a. [Biochemical interrelations between pine and birch]. *Lesnoe Khozyaistvo* (2): 10-12.

Kolesnichenko, M.V. 1963 [Classification of forms of influence of woody plants and some methods of study of questions of allelopathy]. In: *Materialy Nauchnaya Konferentsiya, 1962.* Voronezhskogo Sel'sko-khozyaistvennogo Instituta. Voronezh. Pp. 64-65.

Kolobkova, E.V. and Kudryashova, N.A. 1956. [On inhibitors of germination from seeds of Siberian peashrub]. *Fiziologiya Rastenii* 3(2): 115-120.

Kolobkova, E.V. and Kudryashova, N.A. 1960. [On inhibitors of seed germination]. In: *Issledovaniya po Biologii i Biokhimii Prorastaniya* [=Studies on the biology and biochemistry of seed germination]. Izd-vo Akademii Nauk SSSR, Moscow. Pp. 8-31.

Kolosov, I.I. 1962 *Poglotitel'naya Deyatel'nost' Kornevykh Sistem Rastenii* [=Absorptive activity of the root systems of plants]. Izd-vo Akademii Nauk SSSR, Moscow.

Kolpikov, D.I. 1960. [On the mechanism of self-regulation of extra-stomatal gas exchange in terrestrial higher plants]. In: Referat. I Tezisy n.-i rabot Stavropol'skogo gos. Ped. Instituta vyp. 2. Stavropol'.

Kononova, M.M. 1963. *Organicheskoe Veshchestvo Pochvy. Ego Priroda, Svoistva i Metody Izucheniya* [=Organic substances of soil. Their nature, occurrence and methods of study]. Izd-vo Akademii Nauk SSSR, Moscow. (The second edition of this book was translated into English in 1966, and appeared as *Soil Organic Matter. Its nature, its role in Nature and in soil fertility*, Pergamon Press, London).

Kostychev, S.P. 1933. *Fiziologiya Rastenii* T. 2 [=Plant physiology v. 2]. Sel'khozgiz, Moscow.

Kozhevnikov, A.V. 1950. *Vesna i Osen' v Zhizni Rastenii* [Spring and autumn in the life of plants]. Izdatel'stvo Moskovskogo Obshchestva Ispytatelei Prirody, Moscow.

Krasil'nikov, N.A. 1952. [The role of microorganisms in supplementary plant nutrition]. *Uspekhi Sovremennoi Biologii* 33(3): 321-338.

Krasil'nikov, N.A. 1958. *Mikroorganizmy Pochvy i Visshye Rasteniya* [=Soil microorganisms and higher plants]. Izd-vo Akademii Nauk SSSR, Moscow. (This book was translated into English in 1962, and appeared as *Soil Microorganisms and Higher Plants*, Program for Scientific Translations, Jerusalem).

Kuntze, J.E, and Riker, A.J. 1956. The use of radioactive isotopes to ascertain the role of root in the translocation of water, nutrients and diseases among forest trees. *Proceedings of the International Conference on the Peaceful Uses of Atomic Energy (Geneva)* 12: 144-148.

Kuprevich, V.F. 1949. [Extracellular enzymes from roots of autotrophic higher plants]. *Doklady Akademii Nauk SSSR* 68(5): 953-957.

Kuprevich, V.F. 1951. [Biological activity of soil and methods of its determination]. *Doklady Akademii Nauk SSSR* 79(5): 863-866.

Kursanov, A.L., Krykova, N.N. and Bartapetyan, B.B. 1952. [Movement in the plant of carbon dioxide absorbed through the roots]. *Doklady Akademii Nauk SSSR* 86(4): 913-916.

Kutacek, M. 1963. Glucobrassicin from *Brassica*- and *Raphanus*-species as potential inhibitor of seed germination and root growth. Some remarks on the role of glucobrassicin and ascorbigen in plants. In: *Physiologie, Okologie und Biochemie der Keimung*, Teil II. Ed. H. Borriss. Ernst-Moritz-Arndt-Universität, Greifswald. Pp. 781-786. [1967].

Kutacek, M. Ullman, J. and Liebl, V. 1956. On root excretions. II. The movement of P^{32} in wheat plants grown by the sectional-nutrition method. *Rostlinna Vyroba* 29(6): 525-536.

Kuzin, A.M. and Merenova, V.I. 1953, [On the assimilation of carbon by roots of plants from organic fertilisers]. *Doklady Akademii Nauk SSSR* 90(4): 677-679.

Lasik, J. and Stanek, M. 1963. *Xanthomonas fuscans* (Burkholder) Burk. In the rhizosphere of the common bean (*Phaseolus vulgaris* L.). *Rostlinna Vyroba* 36(7/8): 715-721,

Laštůvka, Z. 1955a. The influence of leaf extracts and root secretions of couch grass on the germination of wheat and rye. *Československa Biologie* 4(2): 103-108.

Laštůvka, Z. 1955b. The influence of couch grass on the growth of wheat and rye. *Československa Biologie* 4(3): 163-174.

Laštůvka, Z. 1955c. The influence of couch grass on glycide and nitrogen metabolism of wheat and rye. *Československa Biologie* 4(7): 422-428.

Laštuvka, Z. and Knoz, J. 1961. *Vliv učinnych latek pyru plaziveho na anatomii pšenice*. Kandidat Dizertacia Prace, Brno.

Lavrenko, E.M. 1959. *Polevaya Geobotanika. T. 1 Osnovnyye Zakonomernosti Rastitel'nykh Soobshchestv i Puti ikh Izucheniya*. [=Field geobotany. Vol. 1. Basic principles of plant communities and ways of study.]. Izd-vo Akademii Nauk SSSR, Moscow.

Lavrenko, E.M., Andreev, V.N. and Leont'ev, V.L. 1955. [Profile of productivity of aboveground parts of natural vegetation]. *Botanicheskii Zhurnal* 40(3): 415-419.

Lavrova, K.G. 1961. [The selection of plants and the potential of its use]. *Uchenye Zapiski Karel'skogo Pedagogicheskogo Instituta* 11(2): 78-81.

Lesik, F.L. and Eviglyanich, M.K. 1961. [Research on the interspecific relationships of plants]. {in Ukrainian]. *Naukovi zapysky naukova-doslidzhennya instytutu pedahohiky, Seryya biolohiya* 16(11).

Levina, R.E. 1957. *Sposoby Rasprostraneniya Plodov i Semyan* [=Methods for propagation of fruits and seeds]. Izd-vo Moskovskogo Universiteta, Moscow.

Linskens, N. F. and Knapp, R. 1955. Über die Ausscheidung von Aminosäure in reinem und gemischten Bestanden verschiedener Pflanzenarten. *Planta* 45(1): 106-117.

Litvinenko, L.G. 1960. [Effect of organic matter on the oxygen metabolism of Black Sea algae]. *Fiziologiya Rastenii* 7(4): 459-466.

Long, W.C., Sweet, D.V. and Tukey, H.B. 1956. Loss of nutrients from plant foliage by leaching as indicated by radioisotopes. *Science* 123(3206): 1039-1040.

Lundergårdh, H. 1957. Klima und Boden in ihrer Wirkung auf das Pflanzenleben. Gustav Fischer, Jena.

Lundergårdh, H. and Stenlid, G. 1944. On the exudation of nucleotides and flavanone from living roots. *Arkiv för Botanik*. 31 (10): 1-27.

Lütthe, U. 1962. Über die Zusammensetzung des Nektars und der Mechanismus seiner Sekretion. Mitteilung III.. Die Rolle der Rüresorption und der spezifischen Zuckersekretion. *Planta* 59(2): 175-194.

Luttkus, K. and Bötticher, R. 1939. Über die Ausscheidung von Aschenstoff durch die Wurzeln. 1. *Planta* 29(3): 325-340.

Madaus, G. 1937. Experimenteller Nachweis der wachstumförderden Wirkung von Blattausscheidungen. *Madaus Jahresbericht* 1: 39-41.

Maksimovich, A.E. 1957. [On the role of the pericarp in balls of beet seed on germination and germ nutrition]. *Trudy Vsesoiuznogo Nauchno-issledovatel'skogo Institut Sakharnoi Svekly* 35.

Manorik, A.V. 1958. [Features of the use of carbon and sulphur from organic and mineral forms of their compounds]. In: *Teziya Konferentsiya po Fiziologiya Rastenii*. Izd-vo Ukrainskoi Akademii Sel'sko-khoziaistvennykh Nauk , Kiev.

Manorik, A.V. and Vasil'chenko, V.F. 1964. [Biologically active substances of organic fertilizers and their significance in plant nutrition]. *Agrobiologiya* (1): 16-28.

Markov, M.V.O. 1958. [On the dominants of phytocenoses in the works of Soviet geobotanists]. *Botanicheskii Zhurnal* 43(4): 504-517.

Markov, M.V.O. 1962. [On the interrelationships among plants in plant associations]. In: Problemy Vnutrividovykh Otnoshenij Organizmov Materialy k Soveshchaniyu po Problemam Vnutrividovykh Otnoshenij Organizmov, 10-14 sentyabrya 1962 Izd-vo

Tomsk Gosudarstbennogo Universiteta im. V. V. Kuybysheva; [Ed.. B. G. Ioganzena] Tomsk. Pp. 59-61.

Markov, M.V.O. 1964. [On the interrelationships among plants in plant associations]. In: *Vzaimootnosheniya Rastenij v Rastitel'nom Soobshchestve*. Izd-vo Kazanskij Gosudarstvennii Universitet im V.I. Ul'yanova-Lenina, Kazan'. Pp. 3-29.

Martin, P. 1956. Qualitative und quantitative Untersuchungen über die Ausscheidung organischer Verbindungen aus den Keimwurzeln des Hafers (*Avena sativa* L.). *Naturwissenschaften* 43(10): 227-228.

Martin, P. 1957. Die Abgabe von organischen Verbindungen insbesondere von Scopoletin aus den Keimwurzeln des Hafers. *Zeitschrift für Botanik* 45(6): 475-506.

Martin, P. 1963. Zur Frage der Abgabe organischer Verbindungen aus Pflanzen-wurzeln. *Zeitschrift für Botanik* 51(3): 300-307.

Michael, G. and Marschner, H. 1958. Untersuchungen über die Phosphat-Abscheidung aus Pflanzenwurzeln mit Hilfe von P32. *Zeitschrift für Pflanzenernährung, Dungung, Bodenkunde* 80(1): 1-18.

Meshkov, M.V. 1952. [Substances that activate the growth of microorganisms in the root excretions of plants]. *Zhurnal Obshchej Biologii* 13(1): 76-86.

Mirchink, T.G. and Greshnykh, K.P. 1961. [The toxicity of fungi of the genus *Penicillium* in relation to plants and animals]. In: *Trudy Vtorogo Soveshchaniya.po Mikotoksikozam, Cheloveka i Zhivotnykh*. Naukova Dumka, Kiev.

Mishustin, E.N. 1961. *Mikroorganizmy i Effektivnoe Plodorodie Pochv* [=Microorganisms and effective fertility of the soil]. Izd-vo Akademiya Nauk SSSR, Moscow.

Mishustin, E.N. 1963. {Determination of the biological activity of the soil] *Mikrobiologiya* 32(3): 911-917.

Mishustin, E.N. and Naumova, A.N. 1955. [Secretion of toxic substances by alfalfa and their effect on cotton and the soil microflora]. *Izvestiya Akademiya Nauk SSSR, Seriya Biologicheskaya* 1955(6): 3-9.

Misnik, G, E. 1949. *Proizvodstvennaya Kharakteristika Semyan Derev'yev i Kustarnikov Gorodskikh Nasazhdeniy* [=Production characteristics of seeds of trees and shrubs of urban spaces]. Izd-vo Ministerstvo Kommunal'nogo Khozyaistva RSFSR, Moscow.

Mitchell, J.W., Linder, P.J., and Robinson, M.B. 1961. Mechanism of root exudation of α-methoxyphenylacetic acid in the bean plant. *Botanical Gazette* 123(2): 134-137.

Mohill, A. and Paleg, L. 1963. The effects of gibberellic acid and sugars on the dark germination of lettuce seeds. In: *Physiologie, Okologie und Biochemie der Keimung*, Teil I. Ed. H. Borriss. Ernst-Moritz-Arndt-Universität, Greifswald. Pp. 249-253 [1967].

Molisch, H. 1937. Der Einfluss einer Pflanze auf die Andere – Allelopathie. Gustav Fischer, Jena.

Molotovskii, G.Kh. 1949. [The significance of inactivators of growth for dormancy in plants]. *Doklady Akademii Nauk SSSR* 68(2): 405-408.

Moskovets, V.S. 1961. *Gribnaya Flora Rizosfery Lyutserny na Yuge USSR*.[= Fungal flora of the rhizosphere of alfalfa in southern Ukraine]. Autoref. Diss., Kiev.

Mothes, K. 1955. Physiology of alkaloids. *Annual Review of Plant Physiology* 6: 393-342.

Muller, C.H., Muller, W.H. and Haines, B.L. 1963. Volatile growth inhibitors produced by aromatic shrubs. *Science* 143(3605): 471-473.

Muromtsev, G.S. and Pen'kov, L.A. 1962. *Gibberelliny* [=Gibberellins]. Izd-vo Sel'skokhozyajstvennoi Literatury, Zhurnalov i Plakatov, Moscow.

Nasonov, D.N. and Aleksandrov, V.Ya. 1940. *Reaktsiya Zhivogo Veshchestva na Vneshniye Vozdeistviya* [=The reaction of living matter to external influences]. . Izd-vo Akademii Nauk SSSR, Moscow.

Nétien, G., and Lacharme, J. 1960. Actions comparées des tétracyclines sur cultures *in vitro* de topinambour. *Bulletin de la Société Botanique de France* 107(4-6): 172-174.

Nezgovorov, L.A. 1959. [Influence the pathogenicity of soil on photosynthesis and chlorophyll content during cooling cucumber plants]. *Fiziologiya Rastenii* 6(5): 585-591.

Nezgovorov, L.A. 1963. [On the question about the reasons for the increased pathogenicity of the fungus *Pythium* in cold soil]. *Fiziologiya Rastenii* 10(1): 55-64.

Niemann, E. 1952. Vergleichende Untersuchungen über die Ausscheidung keimungshemmende Stoffe aus Früchte und Samen unter besonderer Berücksichtigung von *Foeniculum vulgare* Mill. *Flora* 139(1): 185-242.

Niggemann, J. 1955a. Beitrag zur Kennzeichnung von Wirkstoffen des Torfes. *Naturwissenschaften* 42(4): 98-99.

Niggemann, J. 1955b. Zur Kennzeichnung von Wirkstoffen des Torfes. *Naturwissenschaften* 42(11): 346-347.

Nikolaev, G.V. 1963. [Movement of phosphorus, calcium and sulphur from one plant to another through their root systems]. *Fiziologiya Rastenii* 10(4): 441-446.

Novotel'nov, N.V. and Ezhov, I.S. 1956. [On the secretion of antibiotic substances from grains in cereal crops]. In: *Trudy Leningradskogo Tekhnologicheskogo Instituta Kholodil'noi Promyshlennosti*, Vypusk 14.

Oertli, J. J. 1962. Loss of boron from plants through guttation. *Soil Science* 94(4): 214-219.

Osvald, H. 1949. Root exudates and seed germination. *Kungliga Lantbrukshögskolans Annaler* 16: 789-796.

Ovcharov, K.E. 1961. [The role of vitamins in the allelopathy of plants]. *Uspekhi Sovremennoi Biologii* 51(1): 50-61.

Ovcharov, K.E. 1964. *Vitaminy Rastenii* [=Vitamins of Plants]. Izd-vo Kolos, Moscow.

Ozaki, K. and Tai, K. 1962. Nitrogen metabolism of paddy rice at heading. II. Nitrogenous constituents of guttation from the ear of boot stage. *Soil Science and Plant Nutrition* 8(4): 10-12.

Parkinson, D. 1955. Liberation of amino-acids by oat seedlings. Nature 176: 33-36.

Patrick, Z.A. 1955. The peach replant problem in Ontario. II. Toxic substances from microbial decomposition products of peach root residues. *Canadian Journal of Botany* 33(4): 461-486.

Pearson, R. and Parkinson, D. 1960. The sites of excretion of ninhydrin-positive substances by broad bean seedlings. *Plant and Soil* 13: 391-396.

Peers, F.G. 1958. Germination inhibitory substances in oat husks. *West African Journal of Biological Chemistry* 2(1): 9-14.

Peterson, N.V. 1961 [Sources that enrich soil enzymes] [Ukrainian]. *Mikrobiolohichnyy Zhurnal* 23(6): 5-11.

Peive, Ya.V. 1961. *Biokhimiya Pochv* [=Biochemistry of the Soil]. Izd-vo Sel'skokhozyajstvennoi Literatury, Zhurnalov i Plakatov, Moscow.

Persidsky, D.J. and Wilde, S.A. 1954. The effect of volatile substances released by soil, humus and composts on the growth of excised roots. *Plant Physiology* 29(5): 484-486.

Petrii, E. and Chrastil, J. 1955. The exosmosis of flavanone from root explantations of *Arachis hypogaea* L. *Folia Biologia*. (Prague) 1: 310-312.

Pidoplichko, M.M., Moskovets',V.S. and Zhdanova, N.M. 1963. [The influence of some fungi from the rhizosphere of maize on its seedlings] [in Ukrainian]. *Mikrobiolohichnyy Zhurnal* 25(6): 38-43.

Pilet, P.-E. and Kobr, M. 1958. Étude de l'inhibition de la croissance radiculaire par le DL-tryptophane. *Berichte der Schweizerischen Botanischen Gesellschaft* 68: 239-248.

Politis, J. 1948. Sur la distribution de l'acide chlorogénique dans la famille des Solanacés et dans les organes de ces plantes. *Comptes Rendus de l'Académie des Sciences* 226: 692-693.

Polyanskaya, L.A. Nosov, A.K. and Ovcharov, K.E. 1963. *Znacheniye Nekotorykh Vitaminov v Zhiznedeyatel'nosti Tonkovoloknistogo Khlopchatnika* [=The value of certain vitamins in the life of staple cotton]. Izd-vo Akademiya Nauk Turkmenskoi SSR, Ashkhabad.

Porutskii, G.V., Luchko, A.S. and Matkovskii, K.I. 1963. [On the content of ethylene hydrocarbons in plant volatiles]. *Fiziologiya Rastenii* 9(4).

Porutskii, G.V. and Matkovskii, K.I. 1963. [Influence of some chemical stimulants on release of volatile substances (air vitamins) by plants] [in Bulgarian]. *Izvestiya na Institut po Biologiya Metodii Popov, Sofia* 13: 147.

Preston, W.H. Jr., Mitchell, J.W. and Reeve, W. 1954. Movement of alpha-methoxyphenylacetic acid from one plant to another through their root systems. *Science* 119(3092): 437-438.

Pryanishnikov, D.N. 1951. [Absorption and excretion of ammonia by plant roots.]. In: *Izbrannye Sochineniya, T. 1.* Izd-vo Akademii Nauk SSSR. (This article appeared originally in 1929 in *Doklady Akademii Nauk SSSR* (14): 327-331.

Putrimas, A.D. 1963. [Physiological action of phytoncides of the anemonin type]. Lietuvos TSR Mokslu Akad Darbai Ser C 1963:123-140.

Rabinowitch, E. 1951. *Fotosintez* [=Photosynthesis]. Izd-vo Inostrannoi Literatury, Moscow.

Rabotnov, T.A. 1959. [On the work of Ellenberg on the causal study of riparian vegetation]. *Botanicheskii Zhurnal* 44(5): 634-638.

Rachkov, L.D. 1963. [Studies on the role of plants in the migration of phosphorus-32 using components in models of phytocenoses]. *Tematicheskij Sbornik, Otdelenie Fiziologii i Biofiziki Rastenij Akademiya Nauk Tadzhiskoj SSR* 1962(2): 57-65.

Rachkov, L.D. 1963. *Izucheniye Roli Rasteniy v Migratsii Radioizotopov po Komponentam Estestvennykh Biogeotsenokhov i Model'nykh sistem*. [=Study of the role of plants in the migration of radionuclides in the components of natural biogeocenoses and model systems]. Autoref Diss., Sverdlovsk.

Rademacher, B. 1959. Gegenseitige Beeinflusung höherer Pflanzen. In: *Encyclopedia of Plant Physiology. Volume 11. Heterotrophy*, Ed. W. Ruhland: Springer Verlag, Berlin. Pp. 655-706.

Rakhteenko, I.N. 1958 [On the seasonal cycle of the absorption and excretion of mineral nutrients by roots of trees]. *Fiziologiya Rastenii* 5(5): 63-84.

Rakhteenko, I.N. 1963. *Rost i Vzaimodeystviye Kornevykh Sistem Drevesnykh Rastenij* [=Growth and Interactions of Root Systems of Woody Plants]. Akademiya Nauk Belorusskoyu SSR, Minsk.

Ratner, E. I. 1948. [On the status of potassium in plant tissue]. *Doklady Akademii Nauk SSSR* 62(3): 413-416.

Ratner, E.I., and Samojlova, S.A. 1955. [Extracellular phospatase activity of roots]. *Fiziologiya Rastenii* 2(1): 30-41.

Ratner, E.I., Smirnov, A.M., Khuan, Kh-shu, Ukhina, S.F. and Kuzovkina, I.N. 1963. [The assimilation of amino acids as a nitrogen source by isolated roots of lucerne and whole pea plants in sterile culture]. *Fiziologiya Rastenii* 10(6): 673-681.

Ratner, E.I. and Ukhina, S.F. 1963. [Course of transformation of amino acids absorbed externally in roots of maize]. *Fiziologiya Rastenii* 10(4): 393-399.

Ronsal', G.A. 1963. [On the biological activity of soluble humates extracted from compost] [in Ukrainian]. In: Fizioloho - Biokhimichni. Osnovy Pidvyshchennya Produktyvnosti Roslyn [=Physiologo-biochemical bases of improving productivity of plants]. Derzhsil'hospvydav URSR, Kiev. Pp. 67-70.

Rovira, A.D. 1956. Plant root excretion in relation to the rhizosphere effect. 1. The nature of root exudate from oats and peas. *Plant and Soil* 7(2): 178-194.

Rubin, B.A. 1960. *Biokhimya i Fiziologiya Immuniteta Rastenij* [=Biochemistry and Physiology of Immunity in Plants]. Izd-vo Akademiya Nauk SSSR, Moscow.

Rubtsov, I.N. 1950. [New data on the intergrowth of the root systems of some forest species]. *Agrobiologiya* (6): 57-65.

Rudakov, I.F. 1956. [Phytoncides of clematis]. *Uchenye Zapiski* Orlovskogo gosudarstvennogo *Pedagogicheskogo Instituta, Seriya estestvoznaniya i khimii* 10(3): 30-51.

Ruddat, M., Lang, A. and Mossetig, E. 1963. Gibberellin activity of steviol, a plant terpenoid. *Naturwissenschaften* 50(1): 23.

Sabinin, D.A. 1955. *Fiziologicheskie Osnovy Pitanii Rastenij* [=Physiological Principles of Plant Nutrition]. Izd-vo Akademiya Nauk SSSR, Moscow.

Sacher, J.A. 1962. Relations between changes in membrane permeability and the climacteric in banana and avocado. *Nature* 195(4841): 577-578.

Sadulaev, A and Savchenko, M.G. 1964. [Significance of root secretions in the phosphorus nutrition of corn and beans in their joint growth]. [in Ukrainian].In: *Druha Ukrayins'ka Respublikans'ka Naukova Konferentsiya z Fiziolohiyi ta Biokhimiyi Roslyn..* Vid-vo Akademiya Nauk URSR. Kiev.

Samtsevich, S.A. and Borisova, V.N. 1963. [The toxicity of volatile substances produced by microorganisms in the soil]. *Mikrobiologiya* 32(3): 484-491.

Sanadze, G.A. 1959. [Excretion of volatile organic substances by plants]. *Soobshcheniya Akademiya Nauk GSSR* 22: 449-454.

Sanadze, G.A. 1961. *Vydelenie Rasteniyami Letuchikh Organicheskikh Veshchestv* [=Excretion of Volatile Organic Substances by Plants]. Izd-vo Akademii Nauk Gruzinskoj SSR, Tbilisi.

Sanadze, G.A, 1964. [On the conditions for the excretion of the diene C_5H_8 (isoprene) from leaves]. *Fiziologiya Rastenii* 11(1): 39-43.

Saussure, N.T. de 1804. *Recherches Chimiques sur la Végétation*. V. Nyon, Paris

Savulescu, A., Stanescu, N. and Constantinescu, O. 1963. Investigation on the penetration and translocation of certain antibiotics in tomato plants. *Revue de Biologie, Bucharest* 8(4): 409-417.

Scamoni, A. 1955. Einführung in die praktische Vegetationskunde. Gustav Fischer, Berlin.

Scharpenseel, H.W. and Krausse, R. 1962. Aminosäureuntersuchungen an verschiedenen organischen Sedimenten, besonders Grau- und Braunhuminsäure-fraktionen verschiedener Bodentypen (enschliesslich C^{14}-markierter Huminsäure), *Zeitschrift für Pflanzenernährung, Dungung, Bodenkunde* 96(1): 11-34.

Scheffer, F. 1958. Der organisch gebundene Stickstoff des Bodens, seine Verwertbarkeit (auch Harnstoff). In: *Encyclopedia of Plant Physiology. Volume 8. Nitrogen Metabolism*, Ed. W. Ruhland: Springer Verlag, Berlin. Pp. 179-200.

Scheffer, F., Kiskuth, R. and Visser, J.H. 1962. Wurzelausscheidungen höherer Pflanzen. 1. Isolierung und Identifizierung der Wurzelausscheidungen von *Trifolium pratense* L., *Medicago sativa* L., und *Eragrostis curvula* (Schrad.) Nees. *Zeitschrift für Pflanzenernährung, Dungung, Bodenkunde* 97(1): 26-40.

Scheffer, F. and Kloke, A. 1954. Der Einfluss von Antibiotika auf die Entwicklung und den Nahrstoffgehalt von Kulturpflanzen. *Zeitschrift für Pflanzenernährung, Dungung, Bodenkunde* 66(1): 29-38.

Schönbeck, F.E. and Winter, A.G. 1962. Weitere Untersuchungen über die Veränderung der chemischen Zusammensetzung von Pflanzen durch Aufnahme phenol- und cumarinartiger Verbindungen aus Nährlösungen und Boden. *Zeitschrift für Pflanzenernährung, Düngung, Bodenkunde* 98(2): 126-136.

Schwar, C. 1962a. Die Wirkung isolierter Wermutinhaltsstoffe auf *Foeniculum vulgare* Mill., *Lepidium sativum* L. und *Lactuca sativa* L. var. *longifolia* Lam. *Flora* 152(3): 509-512.

Schwär, C. 1962b. Die Wirkung von Wermutextrakt auf den Stoffwechsel des Fenchels (*Foeniculum vulgare* Mill.). *Biologisches Zentralblatt* 81(5) 587-605.

Schwär, C. 1963. Die Wirkung des wermutextraktes auf Keimung, Wachstum und Entwicklung sowie de Wasserhauhalt des Fenchels und andere Pflanzenarten. *Archiv für Gartenbau* 11(2): 91-100.

Segeta, V. 1963. Germination and water absorption of maize caryopses at low temperatures and the exosmosis of sugars as significant factors of cold sensibility. In: *Physiologie, Okologie und Biochemie der Keimung*, Teil I. Ed. H. Borriss. Ernst-Moritz-Arndt-Universität, Greifswald. Pp. 183-190 [1967].

Semergej, K.I. 1951. [Activation of growth and development of cotton by products of the anaerobic decomposition of the roots of alfalfa]. *Doklady Akademii Nauk SSSR* 78(5): 1049-1053.

Shain, S.S. and Kashmanova, V.M. 1959. [The role of penetration into the subsoil by roots of perennial grasses]. *Zemledelie* 1959(11): 56-60.

Shain, S.S., Kashmanova, V.M., Mel;nikova, M.A. and Motova, A.V. 1957. [Sharing of radioactive phosphorus and calcium by forage plants]. *Doklady Vsesoyuznoy Ordina Lenina Akademii Sel'skokhozyaystvennykh Nauk Imeni V.I. Lenina* 1957(1):15-23.

Shain, S.S. and Nikolaev, G.V. 1962. [Some features of the root in plant nutrition]. *Kukuruza* 1962(1): 43.

Shintikov, V.N. 1950. [Spruce sugar] *Priroda* 1950 (11): 88-89.

Sievers, E. 1961. Über antiphytotische Stoffe im Schnittholz der Weinrebe (*Vitis vinifera* L.). *Naturwissenschaften* 48(14): 508.

Sievers, E. 1963. Untersuchungen über die wasserlöslichen Hemmstoffe aus den Schnittholz der Weinrebe (*Vitis vinifera* L.). I. Zur Wirkung der Hemmstoffe auf die Keimung und Entwicklung von Rebsämlingen, *Naturwissenschaften* 50(4): 134.

Sircar, S.M. and Ray, A. 1961a. Growth substances separated from the root of water hyacinth by paper chromatography. *Nature* 190(4782): 1213-1214.

Sircar, S.M. and Chakravarty, R. 1961. The effect of growth regulating substances of the root extract of water hyacinth (*Eichhornia speciosa* Kunth.) on jute (*Corchorus capsularis* Linn.). *Current Science* 30(11): 429-430.

Sitnik, K.M. 1960. [Effect of gibberellin and heteroauxin on metabolism in plants] [Ukrainian]. *Ukrainskii Botanicheskii Zhurnal* 17(6).

Skvortsov, S.S. 1960. [On the nature of some components of plant volatiles]. *Fiziologiya Rastenii* 7(2): 641-649.

Skvortsov, S.S. 1964. *Letuchiye Vydeleniya Rastenii i ikh Rol' v Zashchitnykh Reaktsiyakh protiv Mikroorganizmov* [=Volatile selection of plants and their role in defense reactions against microorganisms]. Dissertation, Pervyi Leningradskii Meditsinskii Institut im. I.P. Pavlova, Leningrad.

Šormova, Z. 1961. Biochemicke aspekty stimulace rostlin. Rozprazy Československe akademie ved, Rada matematickuch a prirodnich ved. 71(9): 1-49.

Sovetov, A.V. 1950. *Izbrannye Sochineniya* (=Collected work). Sel'khozgiz Moscow.

Srivastava, B.I.S. and Meredith, W.O.S. 1962. Mechanism of action of gibberellic acid: inhibition of α-amylase development during germination of barley by chloramphenicol and its reversal by gibberellic acid. *Canadian Journal of Botany* 40(9): 1257-1265.

Stocking, C.R. 1956. Guttation and bleeding. In: *Encyclopedia of Plant Physiology. Volume 3. Water Relations of Plants*, Ed. W. Ruhland: Springer Verlag, Berlin. Pp. 489-502.

Stout, M. and Tolman, N. 1941a. Factors affecting the germination of sugar-beet and other seeds with special reference to the toxic effects of ammonia. *Journal of Agricultural Research* 63: 687-713.

Stout, M. and Tolman, N. 1941b. Interference of ammonia, released from sugar beet seed balls, with laboratory germination tests *Agronomy Journal* 33(1): 65-69.

Stowe, B. B. and Yamaki, T. 1957. The history and physiological action of the gibberellins. *Annual Review of Plant Physiology* 8: 181-216.

Straus, J. 1962. Invertase in cell walls of plant tissue cultures. *Plant Physiology* 37(3): 342-348.

Sukachev, V.N. 1942. [Thoughts on the development of phytocenology]. *Sovetskaya Botanika* 1942(1-3): 5-17.

Sukachev, V.N. 1945. [Biocenology and phytocenology]. *Doklady Akademii Nauk SSSR* 47(6): 447-449.

Sukachev, V.N. 1957. [General principles and a program for study of forest types]. In: *Metodicheskie Ukazaniya k Izucheniyu Tipov Lesa.* Izd-vo Akademii Nauk SSSR, Moscow. Pp. 9-63.

Sverchkov, O.N. 1964. [The effect of volatile aromatic emissions from the plant during flowering on air ionization]. *Ukrainskii Botanicheskii Zhurnal* 21(1): 25-29.

Svinarev, V.I. and Barannikova, E.V. 1956. [The role of the secretions of germinating rice seeds in relations with *Echinochloa crus-galli*]. *Fiziologiya Rastenii* 3(3): 272-276.

Takahashi, N. and Curtis, R.W. 1961. Isolation and characterization of malformin. *Plant Physiology* 36(1): 30-36.

Takijima, Y. 1963. Studies on behaviour of the growth inhibiting substances in paddy soils with special reference to the occurrence of root damage in the peaty paddy fields. *Bulletin of the National Institute of Agricultural Sciences (Japan) Ser. B* 13: 117-252.

Timofeev-Resovskij, N.V. 1962. *Nekotorye Problemy Radiatsionnoj Biogeotsenologii* [=Some Problems of Radiation in Biogeocenology]. Autoref. Dis. Doktora, Sverdlovsk.

Timonin, M. I. 1941. The interaction of higher plants and microorganisms. III. Effect of by-products of plant growth on activity of fungi and actinomycetes. *Soil Science* 52(5): 395-414.

Tokin, B.P. 1960. *Gubiteli Mikrobov - Fitontsidy* [=Destroyers of Microbes-Phytoncides. Sovetskaya Rossiya, Moscow.

Toole, E.H., Hendricks, S,B., Borthwick, H.A. and Toole, V.K. 1956. Physiology of seed germination. *Annual Review of Plant Physiology* 7: 299-324.

Tranquillini, W. 1959. Die Stoffproduktion der Zirbe (*Pinus cembra* L.) an der Waldgrenze während eines Jahres. II. Zuwachs und CO_2-Bilanz. *Planta* 54(2): 130-151.

Treibs, W. 1947. Über das Agropyron. Einen natürlichen aromatischen Kohlenwasserstoff der Queckenwurzel. *Chemische Berichte* 80: 97-101.

Unger, E. 1961. Wystepowanie i rola niektoruch wielofenoli w organizmach roslin wyzszych Occurrence and role of some polyphenols in the bodies of higher plants. *Postepy Biochemii* 7(3): 445-452.

Vancura, V. and Hovadik, A. 1963. The root excretions of plants. *Rostlinna Vyroba* 9(36). 683-687.

Vasil'ev, I.L. 1962. Aeroionizatsiya i Gidroionizatsiya v Meditsine. *Materialy Vsesoyuzn. Konf.- Tashkent*, 1962.

Verderevskij, D.D. 1959. *Immunitet Rasteniy k Parazitarnym Boleznyam.* [=The immunity of plants to parasitic diseases]. Gosudarstvennoye Izdatel'stvo Sel'skokhozyajstvennoy Literatury Moscow.

Vines, H.M. and Wedding, R.T. 1960. Some effects of ammonia on plant metabolism and a possible mechanism for ammonia toxicity. *Plant Physiology* 35(6): 820-825.

Virtanen, A.I 1947. The biology and chemistry of nitrogen fixation. *Biological Reviews* 22: 239-269.

Virtanen, A.I and Hausen, S. von. 1931. Untersuchungen über die Leguminose-Bakterien und –Pflanzen. IX. Die Ausnutzung verschiedener Stickstoffverbindungen sowie des in Wurzelknöllchen.gesamellten Stickstoff e durch Leguminospflanzen. *Biochemische Zeitschrift* 232: 1-14.

Virtanen, A.I. and Laine, T. 1935. Chemical nature of the amino acids excreted by lemguminous root nodules. *Nature* 136:756-757.

Visona, L. and Pesce, E. 1963. Action de quelques antibiotiques sur la symbiose du tréfle rouge (*Trifolium pratense* L.). *Annales de l'Institut Pasteur*105: 368–382.

Vogdarina, A. A. 1961. *Fiziologicheskie Osnovy Dejstviya Insektitsidov na Rastenie* [=Physiological Bases of the Effect of Insecticides on Plants]. Izd-vo Sel'skokhozyajstvennoj Literatury, Zhurnalov i Plakatov, Leningrad.

Vozneseyenskiy 1958 [Absorption of carbon dioxide by plant roots]. *Fiziologiya Rastenij* 5(4):329-335.

Walter, H. 1962. Neue Gesichtspunkte zur Wasserversorgung der Wüstenpflanzen. *Berichte der Deutschen botanischen Gesellschaft* 75(9): 349-350.

Wasicky, R. 1959. Herausschwemmen von Substanzen aus Blättern und seine eventuelle biologische Bedeutung. *Naturwissenschaften* 46(5): 172-173.

Weaver, J.E. 1961. The living network in prairie soils. *Botanical Gazette* 123(1): 16-28.

Welbank, P.J. 1961. A study of the nitrogen and water factors in competition with *Agropyron repens* (L.) Beauv. *Annals of Botany* 25(2). 116-137.

Welbank, P.J. 1962. The effects of competition with *Agropyron repens* and of nitrogen- and of water supply on the nitrogen content of *Impatiens parviflora. Annals of Botany* 26(3): 359-373.

Went, F.W. 1962. Thunderstorms as related to organic matter in the atmosphere. *Proceedings of the National Academy of Sciences* 48(3): 309-316.

West, P.M. 1939. Excretion of thiamin and biotin by the roots of higher plants. *Nature* 144: 1050-1051.

Winter, A. and Street, H.E. 1963. A new natural auxin isolated from "staled" root culture medium. *Nature* 108(4887): 1283-1288.

Winter, A.G. 1955. Untersuchungen über Vorkommen und Bedeutung von antimikrobiellen und antiphytotischen Substanzen in natürlichen Boden. *Zeitschrift für Pflanzenernährung, Dungung, Bodenkunde* 69(1-3): 224-233.

Winter, A.G. 1960. Allelopathie als Stoffwanderung und Stoffumwandlung. *Berichte der Deutschen botanischen Gesellschaft* 73(9): 365-379.

Winter, A.G. and Bublitz, W. 1953. Über die keim- und entwicklungshemmende Wirkung der Buchenstreu. *Naturwissenschaften* 40: 416.

Winter, A.G., Peuss, H. and Schönbeck, F. 1959. Untersuchungen über die Aufnahme organischer Substanzen durch die Wurzeln höheren Pflanzen. I. Phenolische Verbindung. *Naturwissenschaften* 46(18): 536-537.

Winter, A.G. and Rümker, R. 1952. Humus und Pflanze. Die Wechselwirkung von Pflanzen und Boden im Licht neuester Forschungen. *Orion* 7(10): 1-5.

Winter, A.G. and Schönbeck, F. 1959a. Zum Wirkstoffkreislauf Pflanze-Tier-Boden. *Zeitschrift für Pflanzenernährung, Dungung, Bodenkunde* 84(1-3): 11-21.

Winter, A.G. and Schönbeck, F. 1959b. Untersuchungen über die Aufnahme organischer Substanzen durch die Wurzeln höheren Pflanzen. II. Phenol. *Naturwissenschaften* 46 (18): 537.

Winter, A.G. and Sievers, E. 1952. Untersuchungen über die Beeinflussung Samenkeimung durch Kaltwasserextrakte aus der Blattstreu verschiedener Gramineen. *Naturwissenschaften* 39: 191-192.

Winter, A.G. and Willeke, L. 1952. Untersuchungen über die Aufnahme von Penicillin unf Streptomycin durch die Wurzeln von *Lepidium sativum* L. und ihre Beständigkeit in natürlichen Boden. *Zeitschrift für Botanik* 40(2): 153-172.

Woods, P.W. 1960. Biological antagonisms due to phytotoxic root exudates. *Botanical Review* 26(4): 546-569.

Yunovidov, A.P. [Fusion of the root systems of pine in the forest]. *Agrobiologiya* 1951 (4): 33-38.

CONTENTS